Coastal Environments

Edited by Yuanzhi Zhang and X. San Liang

Published in London, United Kingdom

IntechOpen

Supporting open minds since 2005

Coastal Environments
http://dx.doi.org/10.5772/intechopen.83243
Edited by Yuanzhi Zhang and X. San Liang

Contributors
Takeshi Hidaka, Sulakhudin, Denah Suswati, Ganiveth Manjarrez Paba, Rosa Baldiris Ávila, María-Victoria Soto, Joselyn Arriagada, Misael Cabello, Che Abd Rahim Mohamed, Nurhanisah Zakri, Hamidul Huq, Tahmid Huq Easher, Rikito Hisamatsu, Oceana Francis, Linqiang Yang, Harrison Togia, Gleb Panteleev, Fausto López-Rodríguez

First published in London, United Kingdom, 2021 by IntechOpen
IntechOpen is the global imprint of INTECHOPEN LIMITED, registered in England and Wales, registration number: 11086078, 5 Princes Gate Court, London, SW7 2QJ, United Kingdom
Printed in Croatia

British Library Cataloguing-in-Publication Data
A catalogue record for this book is available from the British Library

Additional hard and PDF copies can be obtained from orders@intechopen.com

Coastal Environments
Edited by Yuanzhi Zhang and X. San Liang
p. cm.
Print ISBN 978-1-83881-037-5
Online ISBN 978-1-83881-038-2
eBook (PDF) ISBN 978-1-83881-039-9

For EU product safety concerns:
IN TECH d.o.o., Prolaz Marije Krucifikse Kozulić 3, 51000 Rijeka, Croatia, info@intechopen.com or visit our website at intechopen.com.

Meet the editors

Dr. Yuanzhi Zhang is a Professor and Research Fellow on Coastal Environment and Applied Remote Sensing at Nanjing University of Information Science and Technology in China and at the Chinese University of Hong Kong. Dr. Zhang received his Doctor of Science in Technology at Helsinki University of Technology (now Aalto University) in Finland. Dr. Zhang is the author and co-author of 130 peer-reviewed journal articles and 15 books or book chapters. He received the First-Rank Award of the Guangdong Provincial Prize of Science and Technology, China, in 2013 and the Second-Rank Award, ARCA (Actions for Raising Critical Awareness) Prize at the International Symposium 'Environment 2010: Situation and Perspectives for the European Union", Porto, Portugal, in 2003.

Dr. X. San Liang received his Ph.D. from Harvard University, USA. He has since worked at Harvard, Courant Institute, MIT, Second Institute of Oceanography of China, China Institute for Advanced Study. Presently he is Jiangsu Chair Professor at Nanjing Institute of Meteorology, China. Dr. Liang has wide interest in atmosphere-ocean science, applied mathematics, data science, etc. One of his representative contributions, causality analysis, addresses a fundamental problem in science. A rigorous and quantitative causality analysis has been established from first principles, and applied to problems in different disciplines. Another major contribution involves a set of novel theories and methodologies for tackling atmospheric/oceanic problems such as coastal ocean processes, cold air outbreaks, typhoon genesis, storm track dynamics, to name a few.

Contents

Preface

This book aims to present information on coastal environments, including coastal land use and land cover, coastal ecology and vegetation, coastal infrastructure and urbanization, coastal water and pollution, coastal geodynamics or geomorphology, and typhoons and the impact from field measurements and earth observation data. The significance of the book is the integration between different aspects of coastal land use, coastal vegetation, coastal water, and coastal infrastructures or urbanization with their impacts on coastal environments, including coastal pollution and hazards, such as water pollution, red tide, oil spill, etc. Most of the work presented in this book is based on accurate in-situ observations, or remotely sensed data and new technologies in the monitoring of the dynamic change of coastal environments, such as coastal land use and land cover, vegetation, infrastructure and urbanization, water pollution, coastal geodynamics or geomorphology, and typhoons and their impact. This book is expected to serve as a relatively comprehensive reference for coastal researchers, graduate students, as well as policy makers and coastal resource managers.

We thank all the authors who contributed to this book. All reviewers of the book chapters are sincerely thanked for their excellent work providing very helpful comments and timely feedback. Great guidance and support were received from the Publishing Process Manager, Ms. Romina Rovan and the editorial board of IntechOpen. Part of the data used in some papers was obtained from the projects entitled "The Marine Special Program of Jiangsu Province in China (JSZRHYKJ202007)" and the "National Natural Science Foundation of China (U1901215)". The in-situ observations or satellite data from the projects are highly appreciated.

Dr. Yuanzhi Zhang and Dr. X. San Liang
School of Marine Science,
Department of Marine Technologies,
Nanjing University of Information Science and Technology,
Nanjing, China

Section 1

Coastal Management

Chapter 1

Multilevel Management System for Coastal Areas by Network Governance

Takeshi Hidaka

Abstract

Coastal areas are a critical space for humans and other living things and must be properly managed. Coastal management is complicated by the diverse use of coastal areas by diverse individuals. The proposed method is integrated coastal area management that integrates various sectors. Satoumi has been proposed and practiced in Japan and is one of the voluntary coastal zone managements of shallow water. The author proposed a multilevel management system with Satoumi as a basic element. The multilevel management system manages the waters under the jurisdiction of prefectures through multistep efforts such as providing Satoumi, Satoumi networks, and administrative coastal infrastructure. The author also proposed that multistep efforts be bundled with network governance. In this report, after introducing the multistep management system, the efforts of coastal zone management in Shizugawa Bay and the coast of Kagawa Prefecture in Japan were evaluated as a multistep management system and network governance, and the characteristics and issues of each approach were extracted. Furthermore, the effectiveness of multilevel management systems and network governance were examined.

Keywords: coastal area, coastal management, integrated coastal management, Satoumi, network governance

1. Introduction

The coastal area, the sea area near the coastline or the land area, is a critical place for humans and nature. The coastal area is a shallow, accessible space for humans, that provides a calmer and more comfortable environment than inland. In nature, the coastal area is a spawning ground for aquatic products and larvae and a storage and supply site for nutrient salts and has a function of purifying water quality and the like. Such coastal areas should be properly managed [1].

However, because coastal areas are easily used by humans, they are used by various individuals in various ways, and there are various legal systems related to use and management. To properly manage coastal areas, it is necessary to coordinate such diverse uses and various legal systems and bundle them as a whole to achieve the management objectives [2]. Integrated coastal management (ICM) as a type of coastal management is a government-led management system that attempts to use the legal system. In the United States, the pioneer of coastal management, ICM is the center of coastal management, and the ICM system under the Magnuson Patrice Law of the Gulf of California is famous [3]. Partnerships in Environmental

Management for the Seas of East Asia, based in Southeast Asia, is attempting to promote coastal management under the ICM system [4].

By contrast, in Japan, the system for coastal management is not legalized, but projects for comprehensive management and initiatives for voluntary comprehensive management have been promoted. The concept of Satoumi is the basis of these efforts. Satoumi was defined by Yanagi [5] as "a coastal area where biodiversity and productivity have increased due to the addition of human hands." Satoumi development is often voluntarily undertaken by local residents, including fishermen, and regarded as a voluntary coastal management initiative. I considered that a coastal management should be conducted by a combination of voluntary Satoumi development by local residents and public management by administrative agencies and proposed a multilevel management system based on network governance [6].

The purpose of this report is to evaluate the effectiveness of this management system. To do so, after introducing the basic structure of the multilevel management system and Satoumi in Japan, the management centered on aquaculture in Shizugawa Bay and the comprehensive management in Kagawa Prefecture are analyzed as case studies and evaluated from the viewpoint of a network system. Based on these examples, the effectiveness of the proposed system is examined.

2. Coastal management dynamics and multilevel management systems

2.1 Coastal Management Dynamics

The multilevel management system for coastal areas is a normative model proposed as a mechanism for managing coastal areas by combining various subjects and management methods based on the analysis results of precedent cases and management theory [6]. This idea is based on the dynamics of coastal management, where the complexity of use increases because of the expansion of coastal areas (i.e., land waterfront, coasts of municipalities, sea areas of prefectures, and sea areas beyond prefectures) and changes in organizational and management systems. Similarly, Ramkumar et al. have a dynamic view of coastal management. However, this is organized according to the relationship between the time axis and space, which is different from our method of organizing [7].

The coastal management system is expected to differ by the size of the target coastal zone and the complexity of its use. Therefore, the various efforts conducted in Japan (cases in which the author conducted a direct survey) are arranged in **Figure 1** and depend on the two axes of the target space expansion (scale) and the variety of usage. The horizontal axis is divided into Levels 1 to 4 according to the expansion of the space. The vertical axis is divided into three phases, single-phase, dual-phase, and multiphase, reflecting the variety of usage. Usage is relatively simple at Level 1, the narrowest sea level, and the diversity of usage increases as the target sea area expands. Therefore, the position of the example changes from the state in which the land of the fishing village in the lower left of **Figure 1** is used only for fishing, to the state in which the waters beyond the jurisdiction of the prefectures in the upper right of **Figure 1** are used in the most diverse form. The premise of the multilevel management system is that the management system changes accordingly.

2.2 Multilevel management systems

Based on the aforementioned dynamics of coastal management, using the implications from other studies and the analysis results of case management organizations and management methods at each level, based on the creation of Level 1

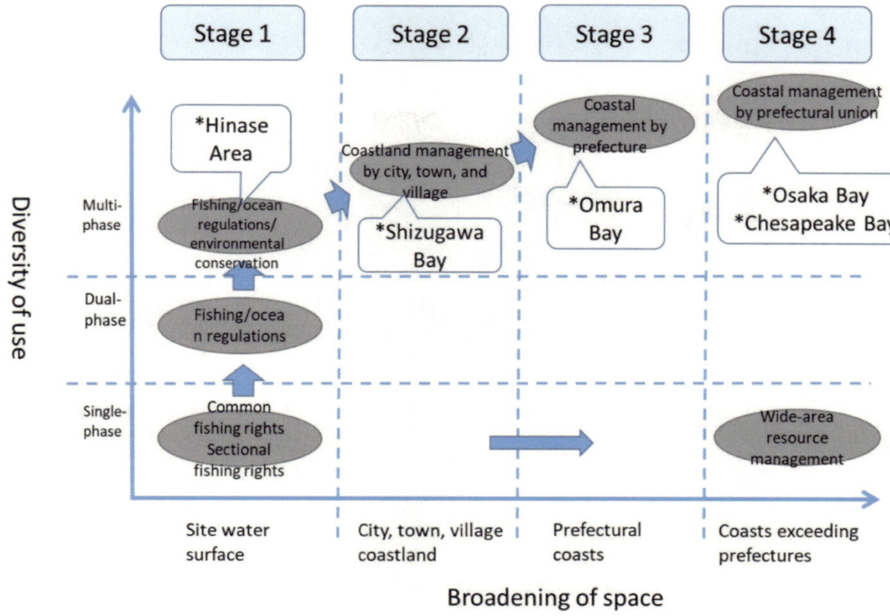

Figure 1.
Coastal management dynamics (modified from Hidaka [6]). Note: "" indicates cases investigated in this study.*

Satoumi, in the Level 2 Satoumi network, we added the provision of coastal infrastructure at Level 3 and the cooperation of marine areas at Level 4, and the combination of these mechanisms was expected to create a coastal management system as a whole: This is the skeleton of a multilevel management system for coastal areas.

At Level 1, the Satoumi project is targeted toward activities where fishermen, users, other local residents, and municipalities are "closely involved in environmental conservation and resource management" in various locations [8]. These activities are conducted by the "whole of regional approach" [9], in which the local and regional individuals involved participate and cooperate.

At Level 2, the Satoumi network is added to refer to the state where the Satoumi projects that are held in various places are organically linked. If Satoumi is networked, a wider coastal area will be covered. For Satoumi to be networked, governments (prefectures, municipalities) and private organizations must support the Satoumi project and the cooperation between Satoumi [10].

At Level 3, the provision of coastal infrastructure (hereinafter, coastal infrastructure) will be added. Coastal infrastructure refers to administrative services provided only by the government, such as regulations for environmental conservation and resource management, and coastal conservation projects. Vertical administration is a problem in the provision of coastal infrastructure, which is efficient because of a "whole of government approach" [11, 12], which is an inter-agency approach.

In other words, at Level 3, Satoumi development, a Satoumi network, and coastal infrastructure is provided collaboratively. The combination of these three efforts at Level 3 enables coastal management in the waters under the jurisdiction of prefectures, which is the basic scope of coastal management.

In a wide area beyond the jurisdiction of the Level 4 prefectures, the related prefectures collaborate through the intermediary of the country to provide coastal infrastructure in common, and Satoumi development and a Satoumi network proceed according to the situation.

The outline of the efforts for each level is shown in **Figure 2**.

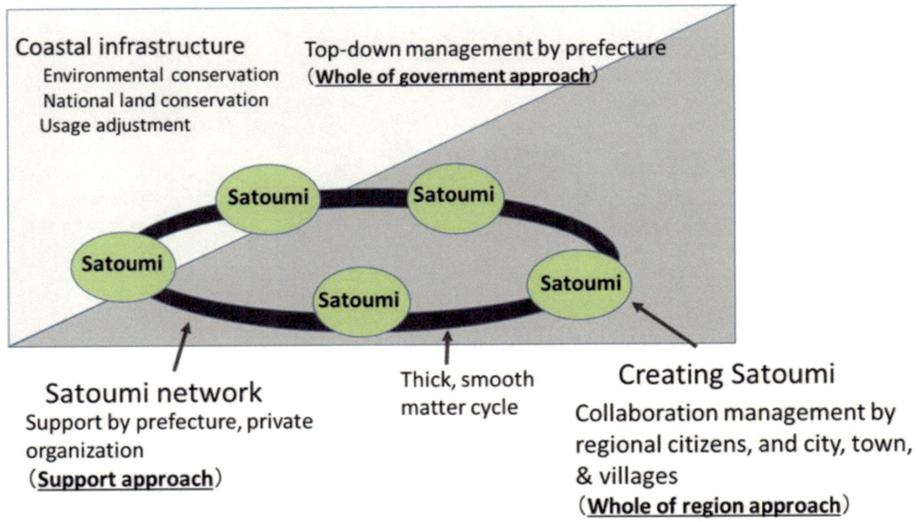

Figure 2.
Basic configuration of the multilevel management system (modified from Hidaka [6]).

The square in the background of **Figure 2** shows the ratio of involvement in management. The management of coastal areas involves the state, prefectures, municipalities, and users. Their roles can be roughly divided into sea area cooperation, provision of coastal area infrastructure, Satoumi network, and Satoumi development. Management bodies are divided into governments (countries, prefectures, cities, towns, and villages) and users (e.g., local residents, fishers), and the decision-making level is hierarchized from a high level that determines policies and systems to the decision-making level of daily activities on site. When arranged side by side, government and private decision-making ratio becomes the rectangle that is the background of **Figure 2**. The left side of the diagonal line is the government's area, and the right side is the user's area; the position of the intersection of the horizontal line and the diagonal line shows the decision-making rate of the government and the user. In other words, the lower the Figure, the higher the user's involvement, the center of the Satoumi project, and its network; the government of the coastal infrastructure is located on the upper side of **Figure 2**.

2.3 Network governance

As described, the multilevel management system comprises a combination of various subjects and a plurality of different managements. Therefore, for this system to operate effectively, a philosophy and process are necessary. Because this management organization is a network type, the idea of network governance was adopted as its philosophy. Network governance is "interfirm coordination that is characterized by organic or informal social system, in contrast to bureaucratic structures within firms and formal relationships between them" [13]. Then, it is explained as "a mode of governance in which stakeholders who possess various resources necessary for governance in the form of a network are integrated, and problem solving is conducted through the process of interaction between those actors" [14]. In other words, a single manager does not govern from the top down; by contrast, the related organizations work together to perform bottom-up governance. In a multilevel management system, the concept of governance is applicable because the contents are not the management by a specific manager but the combination of managements performed by various actors.

Item	Content
Network organization	Formation of network organization with diverse and pluralistic entities involved (composition of council)
Collaboration process	Democratic process involving diverse and pluralistic entities involved (principles of operation of the council)
Equality with the government	Equality role division of government (country, prefecture, municipality), local residents
Multilayer structure that connects different scales	Multilayered organizations and institutions that connect multiple levels, local, national, regional, global (stratification)
Purposing postulate on sustainability	Dropping into the purpose of sustainability criterion, adaptive management process using it

Table 1.
Standards that a multilevel management system should have, based on network governance.

Based on studies of network governance theory and environmental governance theory, five criteria, namely, the network organization, process, equality with the government, multiple layers including different scales, and the aim of the postulate of sustainability were arranged (**Table 1**).

3. Satoumi

The Satoumi is the most basic component of a multilevel management system. The Satoumi proposed by Yanagi is defined as "a coastal area where biodiversity and productivity have increased due to the addition of human hands" [15]. The inspiration for definition is said to be that the forests near the village are maintained by management, such as logging and thinning, by individuals living in the village. In contrast to the conventional belief that no human intervention is required to preserve the marine environment, Dr. Yanagi presented the idea that proper human involvement improves the environment. To prove this idea, Dr. Yanagi searched for cases in various parts of the world and attempted to construct a theory to support it. Examples of demonstrations include the thinning of seaweed beds and traps made of stone walls. The theory is that the biota does not have extreme polarities; instead, there is an optimal relationship between the amount of nutrients and fish resources, and manual maintenance of the appropriate levels of nutrients is required. Recently, the results of a comprehensive investigation into the character of Satoumi under the direction of Dr. Yanagi were reported [16].

The concept of Satoumi proposed by Dr. Yanagi was implemented measures by the Government of Japan in the latter half of the 2000s, for example, the 21st century environmental strategy and the Basic Law of the Sea. Furthermore, the concept of Satoumi was incorporated into coastal management policies by prefectures and municipalities. Notably, the private sector is working on an increasing number of cases to implement Satoumi. According to the author's research, there were approximately 240 cases in 2015 [17], and according to a survey by the Ministry of the Environment, in 2018, there were 291 cases.

Many researchers other than Mr. Yanagi have studied how to manage Satoumi or how to create Satoumi, and the contents of the efforts are evolving, even in the field. Based on these studies, we assess the contents of the Satoumi management that is currently envisioned according to Hidaka [18].

The creation of Satoumi is called Satoumi creation or Satoumi management. The main actors of such activities are local individuals such as fishermen. A council is formed of various local stakeholders, groups, and organizations that centers on fisheries cooperatives and local governments and becomes the management body [17]. For this reason, the conference not only decides the contents of Satoumi management and implements it but also serves as a platform for the participation of various stakeholders and has the role of promoting communication among the participants.

The target of Satoumi management is the shallow area along the coast of the area. This phenomenon is because the individuals involved have the same physical and biological conditions on the land. Instead, the Satoumi cannot cover a large area. Therefore, a Satoumi network must be formed in cooperation with the Satoumi formed in the neighborhood. This network may be formed not only with Satoumi but also with Satoyama, which is woodland close to the village. What is essential is to cover the important points of the material cycle by forming a network. It may also cover the distribution and migration of living things.

In this Satoumi, management activities are conducted, such as preserving and creating the environment of the target sea area, maintaining and increasing fisheries resources, and making rules for use: This is Satoumi management. Unlike conventional fisheries rights management, Satoumi management includes activities not necessarily directly linked to fisheries, and those activities are conducted by applying scientific evidence and making objective observations. The introduction of new technology is also expected. At this time, the existence of scientists closely related to the region is essential because they enable science-based management. In addition, individuals who live outside the area and those who support activities while outside the area are also important players in management. For this reason, the value of the relationship that connects these individuals with Satoumi and the mechanism to realize it is crucial [19].

Additionally, the educational effect of participating in activities as the effect of Satoumi has attracted attention [20]. The literature has observed that not only did the students who participated in the Satoumi project increase their awareness of the region and the environment, but it was also effective for the parents of the students and the fishery personnel who supported the activities.

Furthermore, attention is being paid to the economic effects of Satoumi. From the beginning, an assumption has been that fishery production will be improved by preserving the environment and recovering fishery resources. In addition, educational tourism and marine tourism that participate in the creation of Satoumi are expected to be new economic activities, and a system has been established in which consumers outside the region purchase marine products produced in Satoumi as environmental products. This effort is also to realize the value of the relationships aforementioned.

As mentioned above, the multilevel management system is constructed with the Satoumi as the most basic element. Such Satoumi is formed in the shallow waters of various places, they are networked, and the coastal infrastructure is superposed on Satoumi and their networks to manage coastal areas.

In the following chapters, the cases of Shizugawa Bay, managed in the waters of municipalities, and Kagawa Prefecture, managed in the waters of prefectures, are analyzed from the perspectives of multistep management systems and network governance.

4. Case studies

4.1 Case study multilevel management in Shizugawa Bay

Management of Shizugawa Bay is an example of the practice of Satoumi development and a Satoumi network along the coast of municipalities [21].

Shizugawa Bay is a small open bay on the Sanriku coast of eastern Japan. On the coast of Shizugawa Bay, there are three fishing areas with different characteristics: Tokura area, Shizugawa area, and Utatsu area, where oysters, squirts, coho salmon, and seaweed are cultivated and abalone and sea urchin fisheries are conducted under deferent management ways. Large-scale reconstruction activities are underway in this area because coastal villages and fishing facilities were destroyed by the 2011 Tohoku Earthquake tsunami (**Figure 3**).

Aquaculture in Shizugawa Bay is conducted based on the aquaculture and fishing rights stipulated by the Fisheries Law. The owner of the fishery is the Miyagi Prefectural Fisheries Cooperative. However, a group of fishermen is formed at each of the branches—the Shizugawa branch, Tokura branch, and Utatsu branch—which are the branches of the fishery cooperative, and the actual detailed fishing ground management is performed by each group. This group of fishermen has high character and independence due to differences in the fishing ground environment caused by the historical background and topography of the area where they are based. For this reason, the rules for exercising the fishing grounds are independently decided within each group, and execution management is also performed. This voluntary management of fishing grounds is also important when considering Satoumi management.

Before the earthquake, there were problems such as overcrowding and illegal use of licensed fishing grounds, but after the earthquake, significant improvements have been made in each area. For example, rearranging of aquaculture rafts by the flow of seawater or limiting the total number of rafts to determine the number of rafts per person. In the Tokura area, aquaculture management has been optimized by the drastic reduction of aquaculture density and the introduction of restrictions on the type and amount of aquaculture per capita and has achieved Aquaculture Stewardship Council certification. As a result, the fishing ground environment has been significantly improved in each area, and fishery productivity is also increasing. Based on the aforementioned information, Satoumi is formed in each of the three areas of Shizugawa, Tokura, and Utatsu in Shizugawa Bay and is properly managed for each Satoumi.

In addition, before the Great East Japan Earthquake, because the areas were highly independent, they were managed only by the fishermen, without

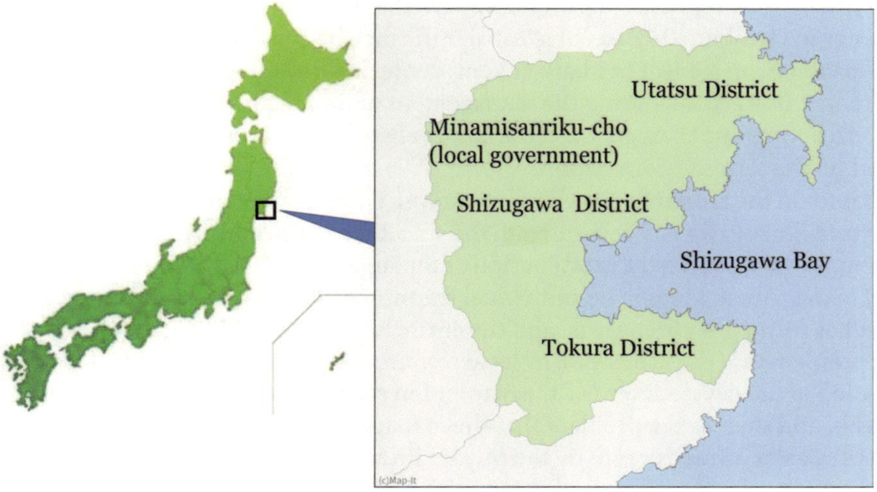

Figure 3.
Location of Shizugawa Bay.

coordination with neighboring areas. However, after the Great East Japan Earthquake, the "Council for Thinking about the Future of Shizugawa Bay" was formed as a consultative organization with various stakeholders and researchers including Minamisanriku Town, a local municipality centered around fishermen. As a result, the management of Shizugawa Bay as a whole is now considered. In addition, a project of the Ministry of the Environment has conducted physicochemical research such as nutrient salt circulation in Shizugawa Bay, and discussions on research studies have been regularly held at the council meetings. As a result, aquaculture management, such as regulation of aquaculture density and facility allocation, has been conducted based on scientific knowledge, and the layout of aquaculture rafts has also been conducted in consideration of the flow of seawater and nutrients throughout the bay. In this manner, although Satoumi in Shizugawa Bay is voluntarily managed by a group of fishermen, it is now managed based on scientific knowledge together with scientists and local stakeholders, and cooperation between adjacent Satoumi is also considered.

By contrast, in Shizugawa Bay, in addition to aquaculture, it is used for sightseeing and nature experiences. Shizugawa Bay has many tourist resources, and approximately 900,000 tourists visit the site each year. Many of the tourism resources are related to fisheries such as recreational fishing, experiential fishing, sightseeing in the bay, and marine-related activities such as scuba diving and sea kayaking, and the demand for marine and environmental education is increasing. Such tourism use of Shizugawa Bay is promoted by the Minamisanriku Town Tourism Association, and many fishermen participate as individuals. However, the management of Shizugawa Bay is not linked to tourism promotion, and there is no organization or plan to consider it.

In addition, Minamisanriku Town has a nature center preparation room (a city agency) for studying conservation and utilization of nature and a visitor center (national and prefecture agency) for introducing the nature and culture of the coast, promotes nature conservation, and has several non-fishery organizations related to Shizugawa Bay, such as nonprofit organizations that support utilization efforts. However, such conservation activities of the natural environment that are not directly related to aquaculture are not linked with the management of Shizugawa Bay.

Notably, although Shizugawa Bay has many regional resources and related organizations, management of Shizugawa Bay is biased toward aquaculture, and a system for comprehensive management of Shizugawa Bay has not been established. To improve this, registration as a seaweed bed in the Ramsar Convention Wetland Register in October 2018 was of great significance because tourism and nature conservation are subject to management, centering on aquaculture and fisheries. Currently, NPOs considering the management of Shizugawa Bay are taking the lead, and efforts are underway toward the comprehensive management of Shizugawa Bay based on the Ramsar Convention.

Based on the aforementioned information, **Table 2** summarizes management efforts in Shizugawa Bay according to the structure of the multistep management system. We summarized the evaluation of Shizugawa Bay management as a multilevel management system, an individual Satoumi centered on aquaculture management has formed in three areas, and the entire bay is managed by networking them. The problem that the utilization of local resources other than aquaculture is not included in the development of the Satoumi in each area and the "regional approach" is weak, and there is the problem that the management of the bay and the provision of coastal infrastructure by the Miyagi Prefectural Government are not linked. Furthermore, the three Satoumi networks have just begun. However, an expectation is that these problems will be resolved by the networking efforts triggered by the registration with the Ramsar Convention.

Management function	Content of management	Management entity
Alliance of prefecture		
Coastal infrastructure	• Basic plan of coast conservation in Miyagi prefecture	Miyagi Prefecture
	• Sanriku Reconstruction National Park	
	• Conservation of wetlands by registration of the Ramsar Convention	
Satoumi network	• Supported by the Nature Center Preparation Office	Minamisanriku-cho
	• Council for enforcement of the Ramsar Convention	
Satoumi	• Fishery utilization plan	Shizukawa branch
	• Effective management of aquaculture at the Shizukawa district, Tokura district and Utatsu district, particularly AFC by Tokura	Togura district
	• Many participants to experience fishing, tourism fishery	Utatsu district
		Tourist Association

Table 2.
Evaluation of Shizugawa Bay management as a multilevel management system.

Next, the management of Shizugawa Bay was evaluated based on the five criteria of network governance.

1. Regarding the network organization, Satoumi was traditionally conducted only by the fishermen's group, but recently, local stakeholders and researchers have joined as advisors. Additionally, if the cooperation between Satoumi and the participation of non-fishermen, which is currently underway, is achieved, it will be able to be evaluated as a network organization. It is still in the formation process.

2. The collaborative process plays a role as a forum for related parties to discuss topics in the "Shizugawa Bay Thinking Council." The goal is that this council will be a place for collaborative processes when the parties involved in non-fishery utilization participate.

3. Regarding equality with the government, because this Satoumi project is based on fishing rights, the branch of the fishery cooperative, which is the subject of rights, and the local municipality, Minamisanriku Town, are equal. However, the relationship with Miyagi Prefecture, which provides coastal infrastructure, is unclear, and equality cannot be evaluated.

4. Regarding the multilayered structure including different scales, we divided it into two stages: Satoumi, which manages each area, and a Satoumi network, which manages the entire Shizugawa Bay. By contrast, seawater and nutrients are flowing into the Shizugawa Bay from the waters outside the bay, and cooperation with the waters outside the bay should be considered. However, this area is off the coastal area covered by Satoumi, which is a management area of Miyagi Prefecture. Collaboration with Miyagi Prefecture should be but has not been fully considered in terms of multilayering.

5. Regarding the purpose of the sustainability standard, the viewpoint that Shizugawa Bay as a whole should aim for this is not defined; thus, this viewpoint

is absent. This absence is because the management of Shizugawa Bay is an accumulation of individual Satoumi management. Currently, the "Council to think about Shizugawa Bay" is working to establish cooperation between the Satoumi, and with the registration of the Ramsar Convention, the entire Shizugawa Bay has begun to be discussed. If this discussion goes well, the expectation is that the sustainability standard will be considered in this criterion.

As described, when the management of Shizugawa Bay was evaluated as network governance, we observed that standards 4 and 5 have not been achieved. The reason for standard 4 is that although the management of the coasts of municipalities has been implemented, cooperation with wider sea areas is necessary even when considering the coasts of municipalities. As for standard 5, there are various uses even in a narrow sea area; thus, sharing the purpose of cooperation of stakeholders is necessary. An expectation is that this will progress in response to the Ramsar Convention.

4.2 Case study multilevel management in Kagawa Prefecture's coast

Management in Kagawa Prefecture was used as an example of the development of Satoumi, a Satoumi network, and the provision of coastal infrastructure in the administrative waters of prefectures [22]. In Kagawa Prefecture, the coastline faces the Inland Sea, including the two open seas, the Sea of Harima and the Sea of Hiuchi. The prefecture has approximately 100 large and small islands with a total coastline of approximately 722 km, the fourth-longest coastline in Japan after Nagasaki Prefecture. In addition, the prefecture area is long in the east and west along the Seto Inland Sea, and even at the longest in the north and south, it is only approximately 30 km and is closely related to the Seto Inland Sea.

Environmental changes in the Seto Inland Sea occurred during the period of high economic growth in the 1960s and 1970s. During that period, industrial development and population growth increased the inflow of drainage from land, increased landfill on the sea surface, and caused the disappearance and modification of shallow water. As a result, the quality of water and the occurrence of red tide caused substantial damage to organisms. To solve this problem, the Seto Inland Sea Environmental Conservation Temporary Measures Act was enacted in 1973; in 1978, it became a permanent law as the Seto Inland Sea Environmental Conservation Special Measures Law (hereinafter referred to as the Setouchi Law).

Regarding the current coastal environment, Kagawa Prefecture has the following five problems (Kagawa Prefecture 2013) [23]: (1) organic pollution that does not show any improvement (Chemical Oxygen Demand environmental standard achievement rate of 40%), (2) disruption of the circulation balance of nutrient salts (occurrence of discoloration of cultured seaweed), (3) seagrass beds that tend to increase but are still small, (4) sea litter that requires immediate attention, and (5) the diminishing relationship between individuals and the sea (**Figure 4**).

In 2012, Kagawa Prefecture made Satoumi development a pillar of its environmental policy in the Seto Inland Sea and added it as a priority policy in 2013. Based on this policy, in April 2013, the "Kagawa 'Satoumi' making council" (hereinafter "council") was established to cover group organizations that may be related to Satoumi and coastal zone management. Since then, this council and its working groups have played an important role. In the council, the working group led the creation of the "Kagawa "Satoumi" vision (hereinafter "vision"), adopted as a pillar of Kagawa Prefecture's environmental policy in Setouchi. Since then, various projects and activities have been implemented based on this vision.

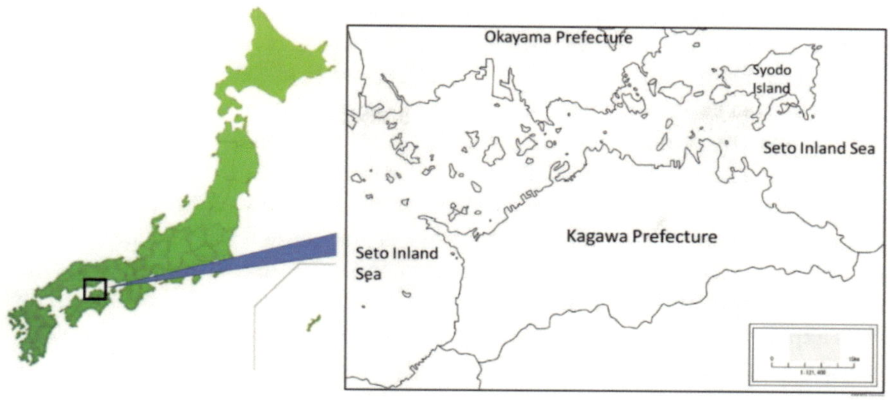

Figure 4.
Location of Kagawa prefecture and the coastline.

The basic idea of the vision is to create "a sustainable and rich sea where individuals and nature coexist." The council and the prefecture call it Satoumi-zukuri. Additionally, as a basic policy of the activity, "the whole prefecture connects all the citizens of the prefecture to connect the mountain, the river, the town, and the sea." In other words, the purpose of the activity policy is to link various stakeholders, activities, and networks.

Based on this basic philosophy and activity policy, to promote initiatives, six points were set: (1) construction of a promotion system, (2) prosperity of common principles and efforts, (3) fostering awareness, (4) human resource development, (5) networking, and (6) adaptive management based on data. Various businesses are attached along these points.

The distinctive feature of the vision-based business is that it is not providing infrastructures that the prefecture directly implements, such as a coastal conservation facility construction business, but practices and ingenuities such as an event or program involving the related organization and prefectural individuals. This measure includes projects that support such software projects. In the activity policy, this will enable all the citizens of the prefecture to conduct activities to use the sea and to conserve it, and to connect businesses that were previously individually conducted in the mountains, rivers, villages, and the sea. For example, in the sea litter reduction project, not only other departments in the prefectural office related to sea litter but also basin municipalities and coastal municipalities have implemented measures cooperatively, for example, activities such as supporting the creation of Satoumi in the area and with related parties.

Regarding the measures that the prefecture should implement as a project under direct control or as a supervisor (not a supporter), we referred to the "Kagawa Prefecture Plan for Environmental Preservation of the Seto Inland Sea" (hereinafter referred to as Kagawa Prefecture Plan), formulated based on the Setouchi Act. This plan summarizes the matters to be performed by the departments of the prefectural office. Regarding the promotion, we supposed that a study of the prefectural office would be across departments. The contents of the Kagawa Prefecture Plan correspond to the coastal area infrastructure and are being tackled as a "whole government approach" beyond the boundaries of departments within the prefectural office.

In addition, Kagawa Prefecture incorporated the vision's activity policy and activities into the contents of the Kagawa Prefecture Plan when it was revised in

2016. As a result, the Kagawa Prefecture Plan integrates the hardware business conducted by the prefecture with the software business conducted by groups and citizens or their support projects. In the Kagawa Prefecture Plan, the council continues to play an important role in creating Satoumi, which was the content of the vision, and the prefecture is responsible for other things.

As aforementioned, this council covers all group organizations that may be involved in Satoumi and coastal area management in Kagawa Prefecture, and 21 groups are members. Although the resolution of the council has no legal effect, it represents the agreement of the parties concerned with the creation of Satoumi in Kagawa Prefecture. The council is working groups of experts in specific fields and has experts as advisors. Currently, there are three working groups, Satoumi Vision, Satoumi Experience, and Satoumi Circulation, which support the creation of Satoumi.

In this manner, the promotion system for business and activities based on the vision is a system in which the prefecture and the prefecture collaborate to promote the vision, rather than the prefecture determining the vision and achieving it from the top down, and support the realization and cooperation of prefecture projects. For the region, the prefecture does not promote business but supports the activities onsite through working groups. Thus, the author referred to Kagawa Prefecture's efforts to create Satoumi a supportive or supportive approach to top-down management.

As the authors observed, the management of coastal areas in Kagawa Prefecture has been conducted in collaboration with the private sector and prefectural government (the former vision), as well as the hardware projects and regulation/management efforts that the prefecture should conduct (former Kagawa Prefecture). In other words, the characteristic of coastal area management in Kagawa Prefecture is that it is integrated under the current Kagawa Prefecture Plan; the former is an effort to support the development of Satoumi and the Satoumi network, and the latter is to provide coastal infrastructure. Therefore, multilevel management in the coastal area of Kagawa Prefecture is being constructed in the Kagawa Prefecture Plan. **Table 3** summarizes the contents of multilevel management.

Next, we evaluated network governance.

1. Regarding the network organization, the vision's activity policy is to connect various stakeholders and activities; thus, this is reflected throughout vision's activities. In particular, the council, the center of the vision, has the characteristics of a network organization because it comprises voluntary participants and the members have equal, mutual relationships.

2. In the consultation process, a forum for consultation by various parties is established based on this meeting. In particular, the working group is a forum for discussions on specific themes.

3. Equality with the government is also secured because the old vision formulated by the council and the old Kagawa Prefecture Plan created in Kagawa Prefecture are integrated in the current Kagawa Prefecture Plan.

4. Regarding the multilayered structure including different scales, it has a three-level structure of Satoumi (shallow sea), Satoumi network (coast of municipalities), and coastal sea area (coast of Kagawa Prefecture). However, we cannot call it a multilayered management structure because there are few efforts by Satoumi.

Management function	Content of management	Management entity
Alliance of prefecture	• Basic plan for environmental conservation in the Seto Inland Sea	National Government
Coastal infrastructure	• Kagawa Prefecture Plan for Environmental Conservation in the Seto Inland Sea	Kagawa Prefecture Government
Satoumi network	• Prefectural office study meeting for Kagawa Prefecture Plan	Kagawa Prefecture Government Related municipalities Non-profit organizations
	• "Kagawa's Satoumi Creation" prefectural office study meeting	
	• "Kagawa's Satoumi Creation" council	
	• Kagawa Satoumi University	
	• Satoumi Concierge	
Satoumi	• Ogoshi area base field	Related municipalities Fisher's organizations Non-profit organizations Citizen's groups
	• Miyagawa area basin matching	
	• Citizen participation monitoring (live creature survey)	

Table 3.
Contents of coastal management in the Kagawa coastal area.

5. Regarding the purpose of sustainability standards, although a qualitative catchphrase has been reached regarding the creation of Satoumi, concrete indicators have been set for "beautiful seas" and "sea of diverse biodiversity." However, there is a problem that the indicators show individual conditions such as water quality. Thus, we recommend improving the index to demonstrate the condition of the entire coastal area of Kagawa Prefecture.

Overall, (1) to (3) of the five criteria are satisfied because the vision focuses on these criteria. Although (4) and (5) are inadequate, we posit that these items are working toward their achievement. Therefore, our evaluation is that the structure of network governance is being prepared.

5. Conclusion

The coastal multilevel management system aims to manage prefectural waters by providing Satoumi, Satoumi networks, and coastal infrastructure, with Satoumi as a basic element. In wide-area waters that exceed prefecture waters, management is performed by sharing the provision of coastal infrastructure by prefectures. This system is a normative model constructed from partial precedents and management theory, and it is necessary to repeat empirical analysis to improve the effectiveness of the model.

The cases in this report are Shizugawa Bay and Kagawa Prefecture. In Shizugawa Bay, management is performed mainly in the creation of Satoumi in a small bay on the coast of the municipalities. In the case of Kagawa Prefecture, Satoumi creation, a Satoumi network, and provision of coastal area infrastructure are comprehensively conducted as a prefecture plan in prefecture waters. Both cases have advantages and disadvantages and are imperfect examples of each stage. However, by understanding each approach as a multilevel management system, we evaluated the advantages and disadvantages of the approach, and was able to roughly infer the effectiveness of this system. Ideally, if the efforts toward Satoumi creation and Satoumi networks such as Shizugawa Bay are combined with Kagawa's comprehensive management of public and private activities and hardware and software projects, a multilevel management system in coastal areas of prefectures will be completed. Notably, the analysis of partial precedents must be repeated. As a result, we will be able to confirm the effectiveness of this system.

The proposal by Hidaka [6] was that network governance can be achieved when the aforementioned five criteria are fulfilled. According to this report, network governance has been achieved at a certain level in Kagawa Prefecture. What remains in Kagawa Prefecture is the development of an ideal state that summarizes the state of the entire coastal area and an index that represents it. The effectiveness of network governance is proved when it has been proved that the whole coastal area reaches the ideal state when the network governance is executed. In that sense, we recommend developing an index that represents the state of the entire coastal area in Kagawa Prefecture.

A limitation of this report is that a wide area outside the jurisdiction of prefectures was beyond the scope of our investigation. In Japan, although pioneering projects are underway in, for example, Tokyo Bay [24] and Osaka Bay [25], there are few cases, and the analysis of actual conditions has been delayed. Thus, further research should analyze the content and results of these efforts. In the United States, the Chesapeake Bay is under interstate coastal zone management efforts [26]. Accumulation of such cases is required for further research.

Acknowledgements

This work was supported by JSPS KAKENHI Grant Number JP 18H03432.

Author details

Takeshi Hidaka
Kindai University, Iizuka City, Japan

*Address all correspondence to: hidaka@fuk.kindai.ac.jp

References

[1] Cullinan C. Integrated Coastal Management Law: Establishing and Strengthening National Legal Framework for Integrated Coastal Management. Rome: FAO Legislative Study 93; 2006. 262 p.

[2] Cullinan C. Integrated Coastal Management Law: Establishing and Strengthening National Legal Framework for Integrated Coastal Management. Rome: FAO Legislative Study 93; 2006. 262 p.

[3] Lester C. CZM in California: Successes and Challenges Ahead: Published online: 16 May 2013. 219-244. DOI: 10.1080/08920753.2013.784891

[4] Chua T. The Dynamics of Integrated Coastal Management: Practical Applications in the Sustainable Coastal Development in East Asia. Manila: PEMSEA; 2006. 431 p.

[5] Yanagi T. Preserving the inland sea. Marine Pollution Bulletin: 19. 1998; 51-53.

[6] Hidaka T. An Attempt of multi-level management system of the coastal area based on network governance. Journal of Coastal Zone Studies, 31(3). 2018; 29-40.

[7] Ramkumar M, Menier D, Kumaraswamy K. Coastal zone management during changing climate and rising sea level: transcendence of institutional, geographic, and subject field barriers is the key. In: Ramkumar M, James R A, Menier D, Kumaraswamy K, editor. Coastal Zone Management: Global Perspectives, Regional Processes, Local Issues. Amsterdam: Elsevier; 2019. p. 1-12.

[8] Kakuma S. What is Satoumi? In: Kakuma S, Yanagi T, Sato T, editors. Satoumi's Recommendation: New Relationship between People and the Sea. Tokyo: Bensei Publishing; 2019. p.9-25.

[9] Hidaka T. Satoumi and Coastal Management: Managing Satoumi. Tokyo: Agriculture, Forestry and Statistics Association; 2016. p.174.

[10] Hidaka T. Satoumi and Coastal Management: Managing Satoumi. Tokyo: Agriculture, Forestry and Statistics Association; 2016. p.240.

[11] Australian Public Service Commission. Tackling Wicked Problems: A Public Policy Perspective. https://www.apsc.gov.au/tackling-wicked-problems-public-policy-perspective

[12] Bevir M. Governance: A Very Short Introduction. Hampshire: Oxford University Press; 2012. 132 p.

[13] Candace J, William SH, Stephen P B. A general theory of network governance: exchange conditions and social mechanisms. Academy of Management Review. 22 (4): 1997; pp. 911-945. doi:10.5465/AMR.1997.9711022109.

[14] Rhodes, R A W. Governance and Public Administration. Debating Governance: Authority, Steering, and Governance. J. Pierre. New York: Oxford University Press; 2000. p. 54-90.

[15] Yanagi T. Sato-umi: New Concept for Coastal Sea Management. Tokyo: TERRAPUB; 2006.110 p.

[16] Yanagi T, editor. Integrated Coastal Management in the Japanese SATOUMI: Restoring Estuaries and Bays. Amsterdam: Elsevier; 2019. 241 p.

[17] Hidaka T, Yoshida M. A study of the structure and function of the Sato-umi management organization: preliminary analysis through questioner surveys.

Journal of Coastal Zone Studies: 28(3). 2015; 107-118.

[18] Hidaka T. The Satoumi concept. In: Li Y, Namikawa T, editors. In the Era of Big Change: Essays about Japanese Small-scale Fisheries. TBTI Global Publication Series, St. John's, NL, Canada; 2020. p. 49-56.

[19] Uehara T, Hidaka T, Matsuda O, Sakurai R, Yanagi T, Yoshioka T. Satoumi: Re-connecting people to nature for sustainable use and conservation of coastal zones. People and Nature: 2019; 1-7. 1 Wiley online library.

[20] Sakurai R, Uehara T, Yoshioka T. Students' perceptions of a marine education program at a junior high school in Japan with a specific focus on Satoumi. Environmental Education Research: 2018; DOI: 10.1080/13504622.2018.1436698.

[21] Hidaka T. A Study of multi-level management system for coastal areas by network governance. In: MEDCOAST 19: Proceedings of the Fourteenth International MEDOAST Congress on Coastal and Marine Sciences, Engineering, Management and Conservation, Vol.1. 2019; 23-32.

[22] Hidaka T. Satoumi and Coastal Management: Managing Satoumi. Tokyo: Agriculture, Forestry and Statistics Association; 2016. p.232-246.

[23] Kagawa Prefecture Government. Vision of Kagawa Satoumi Creation. 2013. https://www.pref.kagawa.lg.jp/kankyokanri/satoumi/vision/satoumibijon25.pdf

[24] Nishimoto T. Tokyo bay revitalizing project. Journal of Coastal Zone Studies: 30(4). 2018; 27-33.

[25] Tsuda Y, Miyahara Y. Osaka bay revitalizing project. Journal of Coastal Zone Studies: 30(4). 2018; 34-40.

[26] Chesapeake Bay Program Science, Restoration, Partnership. Governance and Management Framework for the Chesapeake Bay Program Partnership. 2018. https://www.chesapeakebay.net/documents/Chesapeake_Bay_Program_Governance_Document_version_3.0_%28updated_09.13.18%29.pdf. 20200815

Section 2

Coastal Water and Water Pollution

Chapter 2

Coastal Water: Wisdom, Destruction, Conflicts and Contestation – A Case of Southwest Coastal Region of Bangladesh

Hamidul Huq and Tahmid Huq Easher

Abstract

The coastal zone of Bangladesh is full of opportunities and vulnerabilities. Water is the central source of these opportunities and vulnerabilities. Seawater, river water, canal water, floodplain water, wetlands water, pond water is the dominant source of livelihoods of coastal people. The propel of the coastal economy is dependent on this water. But salinity alone creates vulnerability of the economy and people dependent on this water. Tidal surge, storm surge, drainage congestion, waterlogging, saline water aquaculture are the driving forces of water crises. Water crises are the unsolved issues ever despite large scale interventions. A shortage of freshwater suffers coastal people ever in regards to crop production, drinking water, health, aquaculture, and so on. Uncertainties driven by cyclones, river erosion, outsiders' interventions-led consequences are the big challenges of the coastal zone in managing water. Coastal people are challenging, resilient, adaptive, and strong in contestations in managing water resources for their livelihoods. They exploit the opportunities using their ability of reconstruction.

Keywords: salinity, tidal surge, storm surge, water crisis, water logging, uncertainties, adaptive management, water rights, water management, contestation

1. Introduction

Coastal zones refer to areas where land and sea meet. The coastal zone of Bangladesh is delineated in various ways. Drawing upon a five years long empirical research (2001–2006), the three basic natural system processes and events that govern opportunities and vulnerabilities of the coastal zone of Bangladesh are tidal fluctuations; salinities; and cyclone and storm surge risk [1]. Based on these criteria, the boundary of the coastal zone of Bangladesh consists of 19 districts, where around 42 million people of 158.9 total population of Bangladesh [2] live, with a density of 743 people per sq.km, in a land area of 47,201 sq.km, which is 32% of total land area (147,570 sq.km) of Bangladesh [3]. The projected population of the coastal zone in 2050 is 58 million [1]. There is around 34,775 sq.km area of agriculture land, which is 28% of the total agriculture land area (122,954 sq.km) of Bangladesh [2].

IntechOpen

It is widely argued that water scarcity throughout the world will put mounting pressure on one of the most abundant freshwater ecosystems on earth. Like many large water basins, the Great Lakes water tension has already begun [4], and water tension in the Southwest Coastal Region of Bangladesh has been on escalating trend. As long as coastal water used to be managed by the local people using their wisdom, ecosystems of all forms were functioning naturally. Until the introduction of hard civil engineering designed plans (since 1961), the ecosystems of both freshwater and tidal saline water were as active as is it naturally possible in the southwest coastal region of Bangladesh. These structures are popularly known as 'Polder' under the Coastal Embankment Project. Their purpose was to protect the wetlands from saline water intrusion towards allowing farmers to grow rice at least two seasons a year. But these poorly planned water projects inherited issues like water scarcity, crisis, tension and conflicts in the coastal region. To address the issues generated by the immediate previous projects, one after another structural engineered projects were implemented under the policy arguments of the government, which nothing but magnified the issues.

However, water has always been an emotional issue in the region for thousands of years, but the structural engineered-dominant projects have been creating confusion among the different stakeholders – farmers, fishers, environmentalists, sociologists, and many others. Now the question is, are the millions of people living in this region can be freed from these confusions? It is argued (Ibid), though water issues often vexing, the public is obligated to understand them because water is the foundation of the ecosystem that keeps humans alive. But the abundance of freshwater in the region has been converted into scarcity and uncertainties by the influences of engineering structural water projects over the decades. It is important to help the general public bring the water into focus. Attempts are needed to engage the citizen and the young scientists, academics, professionals in this most important challenge/effort to protect the globally significant waters of the respective region for the next century and beyond.

2. Methodology

This article is written using data of the author's fieldwork mostly focused on the southwest coastal region of Bangladeshduring the 1991's post-cyclone period, ICZMP project during 2002–2005, post-cyclone Sidr in 2007, IWRM research project in 2007–2012, peri-urban water security research project in 2013–2015, ESPA-Delta research project in 2014–2017, women in aquaculture research project in 2014, and the author's post-doctoral research project in 2013–2014 in the coastal zone. The author has interviewed nine key informants among academics, NGO leaders, environmental activists, government officials, and journalists. Rigorous consultations of literature were done to complement the findings from the field research.

3. Coastal water resources system

The Coastal Water Resources System is defined as an integrated system, which performs various functions that refer to the capacity to support and control either natural systems such as storage of floods, facilitation of fish migration or assimilation of wastes; or human and economic activities, e.g., supplying water for domestic purposes, or providing navigable conditions in rivers. The coastal water resources system is naturally a productive system that produces goods and services for meeting up human needs as well as for the maintenance of ecosystems.

It has got an extensive range of water bodies including water resources sub-systems, which are an interlinked system of tidal rivers and channels; riverine flood plains including wetlands; intertidal lands along the coast and estuary branches; lakes and man-made ponds; the groundwater aquifer; and the sea [1]. The Bay of Bengal is the reservoir of seawater (saline water) along the Bangladesh coast. It is a northern extended arm of the Indian Ocean. The total area of the Bay of Bengal is about 510,000 sq.km.

The main sources of fresh surface water are the Ganges, Brahmaputra and Upper Meghna. These mighty rivers drain a basin of about 16,550,000 sq.km, which provides more than 92% fresh surface water to the coastal zone of Bangladesh [1]. The Coastal Zone has a capillary network of rivers and channels, most of them under a season-dependent tidal regime with twice daily variations of water levels and salinities. Ponds are common features in the coastal zone of Bangladesh as the reservoir of freshwater. Ponds are manmade and of different size and shape and depth and are used for different purposes like fish culture, household purposes, drinking water.

4. Salinity

Salinity is defined as the salt concentration, e.g., sodium and chloride in water, which is measured in the unit of PSU (practical salinity unit. Generally, the average salinity in the global ocean is 35.5 PSU, while freshwater like rivers or inland lakes has salinity close to 0 PSU. Observation of river salinity in the coastal zone of Bangladesh is around 10 PSU to 30+ PSU [5]. Salinity plays a significant role in the processes of the water resources system in the coastal zone. The landward intrusion of saline water determines its usefulness for drinking, household purposes, irrigation, aquaculture and other purposes. Salinity distribution in the estuary is strongly influenced by seasonal changes. During the monsoon (June through mid-October), salinity in the estuary drops and water becomes almost fresh. Salinity increases forthe rest of the time of the year with the effect of low discharges of freshwater from river Meghna, or due to further penetration of tide into the river system [1].

Salinity increases have also been caused by the effects of human interventions, e.g., upstream withdrawal of water and reducing the size of flood plains, dry season flow of the Ganges River has decreased since the Farakka barrage was built in India. Farakka Barrage is across the River Ganges located in Murshidabad district in the Indian state of West Bengal, roughly 18 kilometers from the border with Bangladesh near Shibganj. Construction was started in 1961 and completed in 1975 at a cost of US$22 million. Operations began on 21 April 1975. The barrage is about 2304 meters long. The Feeder Canal (Farakka) from the barrage to the Bhagirathi-Hooghly River is about 42 km long having 109 sluice gates. The purpose of the barrage is to divert 1800 cubic meters per second (64,000 cu ft./s) of water from the Ganges to the Hooghly River for flushing out the sediment deposition from the Kolkata harbor without the need for regular mechanical dredging. Out of 109 gates, 108 are over the river and the 109thone over the low-lying land in Malda, as a precaution. The Barrage serves water to the Farakka Super Thermal Power Station. There are also sixty small canals, which can divert some water to other destinations for drinking and other purposes [6]. It has been observed that this water diversion generated negative impacts in the downstream such as salinity levels raised, contaminated fisheries, hindered navigation, and posed a threat to water quality and public health [6].

Another driving force of the increasing trend of salinity is 'Polderisation' in the coastal zone. After the devastating flood of 1954 and 1955, the United Nations commissioned an international mission (known as 'Krug Mission') to solve the flood problem of the country. Following recommendations of this mission, the government implemented the Coastal Embankment Project (CEP) during the 1960s, which included the construction of 'Polders' to protect coastal flood plans from saline water intrusion and tidal surge. A Polder includes [earthen] embankment, sluice gates, and canals. Polderisation follows a process of first: construction of embankment/dike around a low lying area; then the construction of sluice gates to regulate water in and out; excavation of canals to keep internal drainage system active, and to replace the water in the reclaim area with freshwater. Empoldering can be carried out in coastal and inland areas such as lakes. Polders are enclosed by dikes to keep out the sea. To prevent the polders from being waterlogged, they are managed by drainage canals and pumps. Pumps and drainage canals are used to drain the area.

However, the 'polders' and subsequent flood control and irrigation projects converted the wetlands to dry land to facilitate the introduction of high-yielding variety rice which requires controlled irrigation. These interventions disconnected the wetlands from the rivers and prevented sediment formation inside the wetlands which gradually caused the drainage congestion of the rivers as the sediments deposited on the river bed and the river bed became higher than the wetlands in the surrounding basins. Nature's reaction against the intervention was already building up, siltation started getting deposited at the water entry point of the sluice gates, and rivers and canals' bed height began to increase, which resulted in water logging for huge areas and salinity in soil and water of all sources increased up to a level that they were unusable.

Diversion of the Ganges water at Farakka has caused increased river salinity in the southwest region of Bangladesh to intrude further inland. Both the coastal polders and the Farakka barrage had contributed to the gradual siltation of the coastal rivers and are the principal factors contributing to the tidal water level extremes. The coastal agriculture, forestry, industry, and drinking water sectors have suffered enormously as a result of salinity changes in recent years [7, 8].

Saltwater shrimp farming contributes increasingly higher salinity in the coastal zone, especially the southwest region of Bangladesh, since the 1990s. During this time there was a high demand for shrimp in the export market. The outside businessmen, in collaboration with political power and partnership with local large landowners, initiated shrimp farming displacing rice cultivation. Over the 10 years, almost a hundred percent polderised flood plains/agriculture fields got transformed into saltwater shrimp farms. This practice of shrimp farming is continuing. The permanent existence of saltwater in the flood plains generated extreme salinity in soil and surface and groundwater. However, surface salinity is relatively high across the coastal zone. It is projected that salinity will increase in river channels. This increase is more pronounced in the central and western regions with implications for agriculture, shrimp farming and local well-being [5].

5. Water ecosystems services

Until the 1960s, there was no 'development intervention' in water resources development in the coastal zone of Bangladesh. Coastal people enjoyed the ecosystem services of water to meet up their needs. Ponds were used as a source of drinking waterand also as the rainwater reservoir that served the villagers around with freshwater round the year. Open water fisheries were highly adequate. Almost every villager caughtenough fish from floodplains, canals, rivers for their consumption. Farmers grew one crop (Rice) a year. They created seasonal earthen dykes to protect their cropland from saline water intrusion and after harvesting, they abolished the dykes. Farmers also grew some other crops like lentil, mastered seeds, etc. in high lands that are free from tidal surge. This environment refers to a statement that the coastal zone is an attractive place to live and work, with more than 500 million people, including 40 million in Bangladesh coastal zone, living in this environment worldwide [9]. The ecosystem services in the coastal zone, until dominant development interventions, provided for and enhanced the well-being of its human populations. Of course, the benefits to society from nature are dependent on biotic and abiotic earth systems and how these systems interact with social-economic and governance structures (ibid).

The following decade of dominant development intervention in the form of polderisation in the coastal zone in the 1960s experienced social-economic and governance structures' interactions with ecosystems services. The central purposes were served – tidal floodplains were protected from tidal surge and saline water intrusion; three crops of rice in a year in polderised flood plains. Food security was ensured. But, the next decades until the present time, the ecosystems, particularly water and land, experienced destructive interactions with social-economic and governance structures by the massive increasing expansions of saline water shrimp aquaculture in the polderised flood plains displacing rice cultivation.

6. Destructions in water ecosystem services

Increasingly massive shrimp aquaculture influenced changes in water and land use - altering agricultural lands into shrimp farms bringing saline water into freshwater fed croplands. Since the 1980s, shrimp aquaculture was started in the *ghers* - *ghers* are shrimp farms surrounded/impounded by earthen dykes, situated by riversides [10]. Two main factors together provided a catalyst to the process of accelerated shrimp farming: strong international market demand and high prices for shrimp product; and it was no longer financially viable to cultivate rice because the polders had become waterlogged due to poor drainage [11, 12].

Changes in government policies made the shrimp business highly lucrative, shrimp took over as the biggest export earner of Bangladesh [12]. The yearly revenue of saltwater shrimp (*Penaeus monodon,* locally known as *Badga*) were high compare to agriculture. The price for 1 kg of shrimp was up to BDT800 ($10), compared to BDT 25 (32 cents) per kg of rice, with much lower labor and input costs for shrimp. Shrimp was widely considered as 'white gold' that would lead to economic growth and the large farmers converted their agricultural land to shrimp aquaculture farms without considering the negative impacts in long run [13, 14]. With this economic incentive, *gher* owners moved their operations inside the polders by taking land on lease from medium and small farmers, applying muscle power and coercion. Against the law, the *gher* owners bring saline water into the polder by breaching the embankment, saltwater (Bagda shrimp) shrimp aquaculture, which was the beginning of the non-reversible loss of ecosystem services other than saltwater shrimp [15].

Although shrimp farming has a significant impact on the economy of Bangladesh, it has high environmental costs, including the destruction of green vegetation, reduction in crop production, especially rice. Shrimp farming has altered the physical, ecological (aquatic and terrestrial), and socio-economic environment.

Over the decades of the 1980s and 1990s and beyond, shrimp farming has emerged as a major industry in Bangladesh, which has impacts on economic, social and environmental dimensions. The increased salinity in water has created good conditions for shrimp cultivation, a practice that is now the main reason for the increasing soil salinity in Bangladesh. The salinity of shrimp cultivating areas is approximately 500% higher than in non-shrimp cultivating areas, which is extremely contradictory to official purposes/objectives of polderisation under the Coastal Embankment project [16].

7. Water crises for agriculture

'*Water, water everywhere, but not usable for agriculture*', pointed by the farmers of Paikgacha of the southwest coastal region dramatically. This is a common situation concerning the availability of freshwater for irrigation. Saltwater aquaculture, waterlogging, storm surge, salinity in groundwater generated water crises for agriculture activities like plowing/tilling the cropland, raising paddy seedlings, etc. Farmers are to use low quality and inadequate water for irrigation, which reduced the crop yield to the extent that the farmers lost interest in cultivating crops because they cannot afford it. It is also a condition that the growth of rice plants decreases with increased salinity in irrigation water. The groundwater is highly affected by salinity and sodium and continuous use of such irrigation water, causes high sodium soils, breaks down the soil structure, and reduces soil aeration and water infiltration [16–19]. Rainwater is the only source for irrigation of *Aman* rice for most farmers. Heavy rain is required to wash out the soil salinity at the beginning of the rainy season. But, in recent years the rainfall pattern has changed. Rainfall has become erratic and there is a decreasing pattern of rain in the early monsoon which is unfriendly to agriculture. The amount of rainfall is decreasing particularly in the pre-monsoon and monsoon periods.

In the past, farmers used canal water for irrigation, which was fresh. But, since the recent past, the canal water cannot be used for irrigation purpose anymore because of its salinity, which is the contribution of saline water shrimp farms. The condition of pond water is also the same. Besides, the ponds and the canals are occupied by the shrimp farm owners through the means of manipulations and merged with shrimp farming. This practice refers to the absence of good governance and practices of *mal-governance* of water resources management and denial of rights to use of water resources for many purposes of the local people.

One alternative source of freshwater is groundwater, which is not easily available in the coastal zone of Bangladesh. The freshwater table is so deep (250–350 meters, is mostly unavailable) [20]; installation costs of a deep tube well are costly, most farmers cannot afford it. Large farmers privately install deep tube well and supply irrigation water to others on payment, which is also expensive for the medium and small farmers and sharecroppers. The consequences of the excessive amount of water pumped up from the ground/aquifer with the amount recharging it increases the entry of saltwater into freshwater aquifers [16, 19, 21].

8. Water and livelihoods

Water is Life. No one can disagree with this discourse, as long as we are respectful of 'water wisdom'. Wisdom here refers to responsibility that uses in multiple senses: responsible use of water resources; reasonableness towards other uses of water; awareness of what our actions and interventions mean to others, particularly the poor and disadvantaged; and responsibility towards future generations, other forms of life and nature [22].

Livelihoods refer to 'poor' people's living. For them, earning bread is a livelihood. Earning to meet up the basic needs (food, cloth, shelter, health care, and education) is livelihoods. The term livelihood is associated or relevant or applicable only for the 'poor people'. It is applicable only in addressing 'needs'. If it is beyond that, meaning fulfilling 'wants', then it refers to economic growth, which in other words 'economic development'. Economic growth and development refers to meeting up 'wants', which are unlimited, endless, and known as man's greed.

The Coastal zone of Bangladesh was once prosperous fisheries and agricultural hub. Freshwater was available; saline water was beneficial because it flows naturally; the forest was full of resources to serve local people: and the villages were rich in having trees of fruits, timbers; households had have cows, chicken and duck. Overall, the ecosystem services were available at a level that served local people's livelihoods. This inspired me to recall Mahatma Gandhi, "Earth provides enough to satisfy every man's needs, but not every man's greed". Water ecosystem services were available in ample quantity – fishers could catch fish from open water enough for their consumption and to sell for earning cash income; other villagers could catch fish enough for their consumption; villagers could collect vegetables of many types from the crop fields for their consumption. Due to sufficient natural siltation, there were enough crops; there were practices of shared cropping, which provided the landless and small farmers to grow rice that was enough for their annual food stock. Rich bio-diversity and natural environment supported livestock. Farmers were depended on each other for their agriculture work, which kept them tightened in collective initiatives. Thus they lived in harmony; there was little space for inequality and limited power exercise between themselves or by external forces; rich bio-diversity and open access to the natural food sources allowed the poor and disadvantaged people to avoid conflicts with landlords or big farmers [23]. The family structure was simple, joint family – everyone worked and earned for the joint family, work between men and women were segregated; the females looked after the household and in addition to that grew vegetables, fruits and took care of livestock adjacent to their household (ibid).

Today, communities face a regional depletion of natural resources including safe drinking water, and struggle to maintain livelihoods. Both natural and polderisation-induced disasters and the effects of climate change place increasing pressure on the region, hindering livelihoods. Over the past 40 years, development interventions made modifications to the natural environment by controlling the tidal water/rivers. But they failed to control storm surge which is a driving force of ecosystems destructions. On top of that, sponsoring shrimp farming displacing rice production, sponsoring aquaculture in rice fields that centralized the controlling of natural resources in hands of the rich and powerful elites; constructions of engineering infrastructures (roads, bridges that improved transportations to do marketing of industrial products to coastal zone), created huge drainage congestions of rivers, canals, channels. The introduction of tube wells and PSF (pond-sand-filter) technology for drinking water supply by displacing the thousands of years of practice of using [protected] ponds as a dependable (sustainable) source of drinking water. These modifications have caused extensive environmental damage to the point where we are today. Livelihoods are under big threats and the natural environment is extremely fragileand under increasing pressure.

Livestock makes vital contributions in the rural livelihoods in respect to both diet (milk and meat) and generation of income. Livestock faces mainly two types of vulnerabilities due to increased shrimp farming: reduced sources of fodder, and increased mortality rates because of salinity. Saltwater shrimp farming occupied state-owned lands where the people grazed their cattle and also reduced the quantity of fodder and other cattle feed. The current number of cattle had decreased significantly compared with the number of cattle before the period of shrimp farming. The poor farmers either sold their livestock or took them outside of shrimp farming areas [24].

The people's practice of conflicting livelihoods that the contestation between saline water and freshwater in the southwest coastal zone in Bangladesh, can be traced in history in the way water has been managed and the way political-economic forces influenced water systems [25].

The unique tidal wetlands of the southwest have always maintained some level of salinity yet the soil remained fertile and rice production was high. It was not until the introduction of the embankment system and subsequent, promotion and proliferation of shrimp farming that salinity became such a serious problem. Today, the southwest faces a development-induced disaster as salinity infiltrates soil and watertablesthreatens crops andkills vegetation. Shrimpfarming perpetuates and increases salinity levels in the region, reducing options for livelihood diversification and day-laboring opportunities. People are now often forced to migrate to cities for work [26].

9. Drinking water

It was in 1987. I went to Patharghata, an offshore island, under the Borguna district located on the southwest border of the Southwest Coastal Region of Bangladesh for a study purpose. I was having a meeting with a youth club. There was a tube-well (suction pump) in front of the club office. I asked for water to drink. The youth leader asked one member to go to his house and bring water for me. He brought water in a jug and offered me a glass of water. I was surprised to see that they did not offer me the tube-well water. I asked them, why not tube-well water? They said it is not drinkable, because it is too salty. I went to the tube well and tested its water and I was extremely shocked by the taste of water, which was too salty. I drank water that they offered me and found a bit different taste that we do find in tap water in the cities. I asked them the source of that water. That was pond water. They told me, people of this island use pond water as drinking water for thousands of years. After the meeting, they brought me to the pond side. A big pond, full of green with coconut trees on the banks, no other trees, and water was looking so clean. This pond is used only as a source of drinking water, no other purpose. Everyone is abided by this unwritten rule, the youth told me.

My second visit to this island was in 2005 for another study purpose. I met the same person, the then youth club leader, and asked him (after we discussed water and sanitation on the island) about that pond which they used to use as a source of drinking water. He answered me, we were just standing on the bank of that pond, where I saw, at that moment, 10 to15 men were taking bath in the pond, few were washing clothes, one man was cleaning his cow on another side, the water looks unclean, and the *ghat* is with concrete steps and platform for villagers convenient for bathing and washing. He showed me some more concrete work, which is the structure of PSF (pond-sand-filter), constructed by a local NGO with funding and technical supports of an international NGO around three years back. Since then the pond is open to all for uses. But the PSF is not working anymore (after working for about two years). So, the pond water is no longer usable for drinking. The only source of drinking water is few Deep Tube Wells, which is far away from many and saline too.

Historically, people in the coastal zone of Bangladesh, especially the Southwest region, all along used to use pond water for drinking. The community collectively excavated the pond deep as the reservoir of rainwater, constructed earthen banks strong and high to protect the pond from saline water intrusion, planted coconut trees on the blinks for shade on the water to keep the water cool. One pond did serve neighboring two-three villages, even more. Zamindars also excavated ponds to supply drinking water for their citizens/people. But with the influences of 'development interventions' of public health programs on the government using ADB/WB loan, since the mid-1980s, a massive shift from surface water to groundwater sources for domestic water supply. Sadly, in around two decades, the situation turned to reverse: availability of safe drinking water reduced because of arsenic poisoning in tube well water, resulted in the dealing with saline groundwater by the people of coastal areas. The availability of safe drinking water is poor for the coastal communities, as fresh groundwater is only available at great depths, if at all [1].

Department of Public Health Engineering (DPHE) of Bangladesh Government, spending loan fund supports from multilateral organizations, especially the Asian Development Bank, the World Bank, IDBand funding supports of DANIDA, UNICEFimplementeda number of water supply projects include installation of Deep Tubewells, 'Pond-Sand-Filter' (PSF) system since the late 1990s. NGOs have been implementing their PSF projects since 2000.

Despite all these initiatives of development interventions in drinking water supply during the last more than four decades, the coastal people are not ensured with safe drinking water supply. Study [27] shows, at least two-third of coastal rural households fell into the water scarcity and the root causes are saline water intrusion, reduction of upstream flow, sea-level rise, disasters, polder, arsenic contamination, shrimp cultivation in brackish water, excessive use of underground water and lack of appropriate aquifer were highly influential for the disturbance of potable water supply. Water scientists [28] argue that uses of deep tube-wells render the aquifers to overdrawing, which is a potential cause for upcoming. The overdrawing of groundwater is also contributing to declining the capacity of freshwater in flushing out the saline water from the aquifers. This is becoming a great concern in this region [28]. It is argued that recharge to deep aquifers is extremely low in southwest coastal Bangladesh. Water at a depth between 100 and 300 m in this area is a few thousand to >10 thousand years old, suggesting that these aquifers are not receiving any current recharge [28–32]. It is so unfortunate, this scientific knowledge of groundwater dynamics is often ignored in the development interventions in the water supply sector in the southwest coastal region of Bangladesh leading to high risks of water shortage and water crises.

Currently, the coastal rural households are dependent on tube well water, which is not saline and arsenic-free; PSF water, which is available only for monsoon months and only where PSF projects were implemented; rainwater that villagers harvest; and open pond water. One study shows, in some cases, 97% of local people collect their drinking water directly from ponds [33]. Ministry of Water Resources of Bangladesh Government has recently initiated a new project of excavation *deeghi* (big pond) in the southwest coastal region to ensure safe drinking water supply for the local people.

The whole experience of development interventions inthe 'coastal water supply' sector of Bangladesh can be denoted as capillary chaos of projects and programs initiated by the outsiders, which generated permanent water crises in the coastal zone. Water crises refer to 'grossly inequitable distribution of the available water; the decline of traditional water management and conservation systems; the disappearance of once-numerous water bodies; the damage to ecological systems from the interventions in nature in the form water resources development projects; the infliction of hardship, inequity and injustice on poor, disadvantaged communities particularly the ethnic groups, and on women; and uncontrollable, unmanageable generation of waste of all kinds, and the consequent reduction in the availability of water' [22].

Drinking water in the southwest region, both surface and groundwater, has become unfit for human consumption since the salinity has exceeded the recommended level of 960 μmho/cm for potable water since 1987 [7, 34]. Drinking water from natural sources in coastal Bangladesh has become contaminated by varying degrees of salinity due to saltwater intrusion from rising sea levels, cyclone and storm surges, and upstream withdrawal of freshwater [13].

In the pre-shrimp farming period, pond water could be used for drinking purposes even in the dry season, but after the introduction of shrimp cultivation, the pond water becomes too salty to use even for bathing in summer. There are deep tube wells, which are privately owned by the large and the middle farmers. The poorer households depend on these deep tube wells for drinking water. In the previous time, the scarcity of drinking water was not as much as it is at present [25]. As per the opinions of the specialists, the main causes of drinking water scarcity are salinity, arsenic, and the shortage of groundwater. The sea level of this region is rising 3–4 ml per year and it creates new salinity affected areas, which creates further scarcity of drinking water [35]. The average estimated sodium intakes from drinking water ranged from 5 to 16 g/day in the dry season, compared with 0.6–1.2 g/day in the rainy season. The average daily sodium excretion in urine was 3.4 g/day (range, 0.4–7.7 g/day). Women who drank shallow tube-well water were more likely to have urine sodium >100 mmol/day than women who drank rainwater [odds ratio (OR) = 2.05; 95% confidence interval (CI), 1.11–3.80]. The annual hospital prevalence of hypertension in pregnancy was higher in the dry season (OR = 12.2%; 95% CI, 9.5–14.8) than in the rainy season (OR = 5.1%; 95% CI, 2.91–7.26). The estimated salt intake from drinking water in this population exceeded recommended limits. The problem of saline intrusion into drinking water has multiple causes and is likely to be exacerbated by climate change-induced sea-level rise [13]. This study finding suggests that the mean sodium intake in pregnant women is well above WHO/FAO–recommended levels and above those of many other countries. Hypertension in pregnancy is associated with increased rates of adverse maternal and fetal outcomes, both acute and long term, including impaired liver function, low platelet count, intrauterine growth retardation, preterm birth, and maternal and prenatal deaths. The adverse outcomes are substantially increased in women who develop superimposed (pre)eclampsia. It further suggestshypertension in pregnancy is associated with increased rates of adverse maternal and fetal outcomes, both acute and long term, including impaired liver function, low platelet count, intrauterine growth retardation, preterm birth, and maternal and prenatal deaths. The adverse outcomes are substantially increased in women who develop superimposed (pre) eclampsia [36].

10. Discussions

Coastal people are naturally resilient to natural hazards. They are educated by nature. They are knowledgeable about the coastal context up to a higher level. Their knowledge is rooted in 'learn by doing'. They are born to win over the challenges of exploiting the opportunities of livelihoods. Their day-to-day life-world is full of risks, threats, pressure along withthrills, joys, and happiness. Philosophy and Forms of their initiatives of exploiting natural resources – ecosystem services of all forms are embedded in fulfilling needs, not wants. They have followed this discourse for thousands of years until the 'development interventions' were introduced on the coast in the recent past (the 1960s). I have discussed these interventions in the previous sections.

Local people, from their full understandings of possible consequences of the proposed development projects, opposed, protested, and non-cooperated the implementations of the projects. For example, while implementing the Coastal Embankment Project (CEP) local people registered their protest against the project identifyingthe wrong design and irrelevance of the project. Violent protests were also there. But the CEP was implemented and contributed dramatic increases in rice production in the embanked/polderised area. Farmers were able to harvest two or even three bumper crops per year. But nature's reaction against the intervention was already building up. Within 15 years of the construction of embankments, siltation started at the water entry point of the sluice gates and rivers and canals' bed height began to increase. As a result, the polderised flood plains started getting water-logged one after another.

The local people first contested one of the projects of the Coastal Embankment Project (CEP): one five-vent sluice gate became nonfunctioning because the link canal got silted up, in six years of its construction in 1967, resulting in waterlogged areas in 1973. People demanded a solution, but no response was there from the government side. People waited for three years but no action was taken. Then local people (in 1976), especially the farmers were organized and excavated an alternative canal (because people have no access to government built structure to do any repairing work) and connect it directly to the link river (*Bhadra* river), which released the waterlogging of 65,000 hectares land and 54 villages. In this case, water-logging is referred by FAO technically to a situation when the level of groundwater meets the plants' root zone [37]. This may last for at least three months and may prolong up to 8–9 months or even become perennial. The depth of flooding varies, according to the topography of the area, and can reach up to 3 m. This study grouped the effects of water-logging into two categories: (a) immediate loss of life, property, and access to essential services such, e.g., potable water and food, humanitarian assistance, and (b) damage to infrastructure and other assets which underpin livelihoods, health, and sanitation, shelters, etc. They further assessed, at the homestead level, the direct impacts of water-logging is the loss of shelter/house, loss of animals, plants, trees, and access to safe food and water. The affected communities are deprived of basic services such as health, children's education. Over the longer term, as water stands and stagnates, health risks go higher. This study suggests that during waterlogged periods both the poverty and nutrition situation quickly worsens, negative coping strategies, e.g., sale of assets, are adopted, that insecurity due to waterlogging may be a factor in early child marriage, and that spread of disease and social breakdown combine to aggravate underlying vulnerability.

Some other studies showed that within 10 years of implementation, the ill effects of the polder surfaced in massive areas such as many drainage canals became inoperative due to siltation, rendering vast tracts of lands waterlogged all year round [38]. The civil engineering structures impeded vast volumes of sediment-laden monsoon flood flows. The floodwaters caused consequently deposited silt and sediment in the riverbeds and channels. The effect caused a reduction in the bulk-carrying capacity of the water by the rivers and channels, leading to further flooding due to severe congestion of drainage, which progressively led to water-logging. It is classed as a man-made disaster. The cumulative impacts were increased salinity, loss of soil fertility, a decrease in income, worsening of sanitation conditions, loss of livelihoods, and problems in gaining access to residents' homes, agricultural land, and infrastructure facilities. Many people were compelled to move onto embankments and roadsides. Educational institutions were severely damaged and remained closed; children were forced to discontinue schooling. Biodiversity and livestock were adversely affected. Safe drinking water became scarce. Waterborne diseases like diarrhea and scabies became epidemic. Moreover, unemployment forced many people to migrate to cities. Strong competition for the rapidly diminishing resource base heightened tensions and conflicts between sectors of society and created a volatile social situation.

However, collective initiatives and actions of local people to address the issues like waterlogging continued. One of the other experiences of contestations occurred in 1986. After 15 years of construction of two parallel large sluice gates on a deep river (Hari river in 1965), the river was silted up and resulted in waterlogging, which flooded 139 villages and croplands around. Local people of all strata demanded the removal of waterlogging but got no response from the govt. side. So, thousands of villagers collectively cut the embankment at an appropriate point that resulted in releasing waterlogging from this area.

Immediate and continuous consequences of the engineering structure dominant 'flood control' projects over water systems in the southwest coastal region of Bangladesh compelled the local people contesting the interventions that work against interests of naturally grown natural systems of ecosystems of all forms, which provide local people with services to meet up their needs. But these outsiders' designed projects, ignoring and undermining the science and wisdom of ecological systems, embrace explicit notions of befitting the outside professionals, businessmen, politicians, and civil bureaucrats both immediately and in long run. These contestations exist since the project's interventions until now for the survival of the local people. These include organized protests, collective actions to solve the issues, and local initiatives of managing ecosystem services for local people's livelihoods and reducing disaster risks. For example, among many collective actions, one action took place in July 1988. More than 20 thousand people were organized and made a 'public cut' of an embankment, released a big shrimp farm from logged saline water, and brought the land back to rice cultivation. The govt. parties engaged hired terrorists and police against the mob, one farmer and policeman were killed. Govt party sued 300 farmers. Another historical people's movement against a system rehabilitation project, which took place in 1990. Knowing the project design/ plan, the local people were convinced that this project would not help in releasing waterlogging in large wetlands, locally known as *BeelDakatia*. The govt. line agency Bangladesh Water Development Board (BWDB) started the project into action on people's protests. At one point the project river dredger got trap by siltation in the river (Solmari river). Mass agitation inoculated against the project, which was eventually withdrawn after completing only 11% of the required construction. Then in September 1990, a large number of people gathered and cut the embankment to release waterloggedBeelDakatiathrough connecting regular tide with the link river (Hamkura river in the area). Through regular tidal actions and the accumulation of alluvium, the land formation process of the *Beel*resumed [39].

Conflicts of disciplinary boundaries, as well as professional knowledge versus local knowledge and people's wisdom, exist in the polderisation processes all along with the project life. Repeated failure of the 'system rehabilitation' approach throughout the 1980s, 1990s, and beyond invoked public protests and collective actions. In cases of implementation of Drainage Rehabilitation Projects, defying army deployment, local people took civil actions that included road blockades, burning cars of the project officials and government high officials visiting project sites; public cut of the embankment to release the stagnant floodwaters and at the same time, to allow tidal inflows to let the natural circulation of water. These contestations worked up to a certain level protecting local interests and popularized in the whole coastal zone and the knowledge world. Following lessons learned and experiences, the local people demanded that their knowledge of 'Tidal River Management' - to allow tidal flow in the basin to increase tidal volume, to store floodwater during flood current and to trap sediment during the long storage

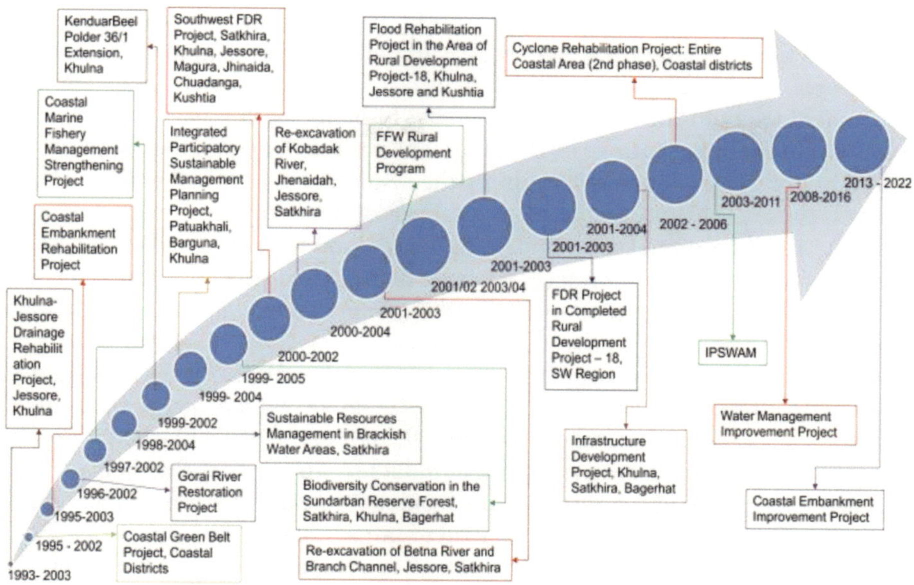

Figure 1.
Structural engineered water projects over the years in the southwest coastal regions of Bangladesh.

period of sedimentation–must be taken into account of project processes, particularly in coastal drainage improvement project design. But the government line agencies keep denying it. Rather the government line agencies and their development partners together have been undertaking projects one after another showing justifications of addressing issues generated by the previous projects. One may easily argue, next projects are taken up by the government to deal with the problems caused by the previous projects but very project generates new problems and escalate the water-related complexities to a further higher level (See the **Figure 1**). The processes of coastal water projects and funding are complex and not easy to understand. Many seek to profit from it and would wish it to continue [37]. But it is expected that the local people's demand together withthe pressure of intellectual work from home and abroad, and negotiations of the civil society remain continued for the interests of protecting coastal zone ecosystems.

11. Conclusion

Historically, the management of coastal water resources was in hands of local people for thousands of years. Water is life, water for life – for humans, animals, trees, fish, biomass, and so on are naturally relevant to coastal overall life-world. Coastal nature is grown naturally and water is the life of this nature. Coastal life-worldis rooted in water systems. Water is the dominant system of overall biophysical systems that embraces major opportunities and vulnerabilities of the coastal people.

Water means to coastal people is surface water – sea, river, canal, pond, *beel*/floodplains, and rainwater. Other than rainwater, all the systems are guided by a unique natural system which is called 'Tidal System'. If any external intervention is designed in technical/engineering science or even social science or multidisciplinary approach, it is essential to understand this tidal water system. Otherwise, you are absolutely wrong, and if you implement any project, which is designed ignoring and undermining the deeply rooted complexities of the tidal water system, you generate water-related problems of all forms in the coastal zone permanently. And it happened to the coastal zone, particularly in the southwest region of Bangladesh.

Water was a natural resource, which required no economic investment for its management for thousands of years in the case of Bangladesh's coastal zone until the 1960s. Heavy interventions by development projects began in the 1960s, which resulted in problems of so many kinds for the insiders that required more projects to address those problems, and implementations of new projects generated further problems, which required further projects. The coastal water has been using as a commodity of development projects business of outsiders- the politicians, businessmen, professionals, multilateral moneylenders, international and national NGOs, consultancy firms, water industries, and so on.

Coastal water is made a 'commodity' from 'natural resource' with the influences of water sector projects. 'Water Resources Management', which was in hands of local people for thousands of years, has been shifted to the 'Water Development' paradigm, which ensured the protection of outsiders' interests at the costs of continued and sustained sufferings of the insiders. Costal water is no longer within the control of insiders, but a central control of outsiders has been already established, which will remain established unless the government draws a hard-line of "Tradeoff".

Author details

Hamidul Huq[1*] and Tahmid Huq Easher[2]

1 Institute of Development Studies and Sustainability (IDSS), United International University (UIU), Dhaka, Bangladesh

2 Department of Environmental Science and Management, North South University, Dhaka, Bangladesh

*Address all correspondence to: hamidulhuq@idss.uiu.ac.bd

References

[1] Islam, M.R. and Ahmad, M (2004), Living in the Coast: Problems, Opportunities and Challenges. PDO-ICZMP, WARPO, Ministry of Water Resources, Government of Bangladesh. Available at: http://www.warpo.gov.bd/rep/liv/living2.pdf. p-33,42,47,130.

[2] BBS (Bangladesh Bureau of Statistics), Statistical Pocketbook - 2015 - Bangladesh Bureau of Statistics. http://203.112.218.65:8008/WebTestApplication/userfiles/Image/LatestReports/PB2015.pdf

[3] BBS (Bangladesh Bureau of Statistics), 2003, Population Census—2001. Community Series. Bangladesh Bureau of Statistics, Ministry of Planning, Dhaka. https://www.scirp.org/(S(lz5mqp453edsnp55rrgjct55))/reference/ReferencesPapers.aspx?ReferenceID=1878157

[4] Annin P. The Great Lakes Water Wars. 1st edition. Island press, Washington DC,2009.

[5] Bricheno, L and Wolf, J (2018), Modelling Tidal River Salinity in Coastal Bangladesh, in Nicholls et al eds., Ecosystem Services for Well-being in Deltas: Integrated Assessment for Policy Analysis, Palgrave Macmillan, Switzerland. p-317,330.

[6] Wikipedia: https://en.wikipedia.org/wiki/Farakka_Barrage

[7] Mirza, M. Q (1998), Diversion of the Ganges Water at Farakka and Its Effects on Salinity in Bangladesh. Environmental Management, Vol. 22, No. 5, pp. 711-722 r 1998 Springer-Verlag New York Inc.

[8] Mondal, M.S., Jalal, M.R., Khan, M.S.A., Kumar, U., Rhman, R., Huq,H (2013), Hydro-Meteorological Trends in Southwest Coastal Bangladesh: Perspectives of Climate Change and Human Interventions, *American Journal of Climate Change*, 2013, 2, 62-70 doi:10.4236/ajcc.2013.21007 Published Online March 2013(http://www.scirp.org/journal/ajcc)

[9] Ericson, J.P., C.J. Vorosmarty, S.L. Dingman, L.G. Ward, and M. Meybeck (2006), Effective sea-level rise and deltas: Causes of Change and human dimension implications. Global and Planetary Change. https//doi.org/10.1016/j.gloplacha.2005.07.004.

[10] Islam, M. S., Milstein, A., Wahab, M. A., Kamal, A. H. M., and Dewan, S (2005),Production and economic return of shrimp aquaculture in coastal ponds of different sizes and with different management regimes. Aquaculture International (2005) 13:489_500 _ Springer 2005. DOI 10.1007/s10499-005-9000-7

[11] Karim, M (1986), Brackishwater shrimp culture demonstration in Bangladesh. agris.fao.org

[12] Alauddin, M. and M. A. Hamid (1999) Shrimp culture in Bangladesh with emphasis on social and economic aspects in Paul Smith (Ed) Towards sustainable shrimp culture in Thailand and the region, ACIAR Proceedings #90, 53-62, Australian Centre for International Agricultural Research, Canberra. Available at: http://aciar.gov.au/files/node/2196/pr90chapter09.pdf

[13] Khan, A. E., Ireson, A., Kovats, S., Mojumder, S. K., Khusru, A., Rahman, A., and Vineis, P (2011), Drinking Water Salinity and Maternal Health in Coastal Bangladesh: Implications of Climate Change. in Environmental Health Perspectives. 2011 September; 119(9): 1328-1332. Published online 2011 April 12. doi: 10.1289/ehp.1002804 PMCID: PMC3230389 Research

[14] Adams, H., Adger, W.N., Huq, H., Rahman, R. and Salehin, M. (2013)

Wellbeing-ecosystem service links: Mechánisms and dynamics in the southwest coastal zone of Bangladesh, ESPA Deltas Working Paper #2, UK, source: www.espadeltas.net.

[15] Nandy, G., Ali, S., and Farid, T.I. (2007) Chingri O Jano-Arthoniti: Kar Lav Kar Khoti (Shrimp and People-Economy: Who gains who losses), ActionAid Bangladesh supported publication (Bangla). Dhaka.

[16] Rahman, M. H., Lund, T. and Bryceson, I (2011), Salinity impacts on agro-biodiversity in three coastal, rural villages of Bangladesh.Ocean & Coastal Management, Volume 54, Issue 6, June 2011, Pages 455-468. Available at: http://www.sciencedirect.com/science/article/pii/S0964569111000317

[17] Miah, G., Bari, N., and Rahman, A (2004), Agricultural Activities and their Impacts on the Ecology and Biodiversity of the Sundarbans area of Bangladesh, in Journal of the National Science Foundation of Sri Lanka>Vol 31, No 1&2 (2003). Available at:http://www.sljol.info/index.php/JNSFSL/article/view/3032

[18] Gain, P., Mannan, M. A., Pal, P. S., Hossain, M. M., and Parvin, S (2004), Effect of salinity on some yield attributes of rice. Pak. J. Biol. Sci., 7 (2004), pp. 760-762

[19] Shamsuddin, S., Xiaoyong, C., and M.K. Hazarika, M. K (2006), Evaluation of groundwater quality for irrigation in Bangladesh using geographic information system. J. Hydrol. Hydromech., 54 (2006)

[20] Datta, D.K. and Ghosh, P.K. (2015), Groundwater of the municipalities of southwestern coastal Bangladesh, in: Subramaniam, V. (Ed.) Surface and Sub-surface Water in Asia: Issues and Perspectives, IOS Press, The Netherlands.

[21] BADC (2004), Survey Report on Irrigation Equipment and Irrigated Area in Boro/2003 Season. Bangladesh Agricultural Development Corporation (BADC). Dhaka

[22] Iyer, R. R (2007), Towards Water Wisdom: Limits, Justice, Harmony. Sage Publications India Private Ltd, New Delhi.p-224,41.

[23] Huq, H (2015), Polderisation in Tidal Floodplains: Exploring Impacts on Social Processes in Bangladesh Southwest Delta. CSD, ULAB, Dhaka

[24] Swapan, M. S. H. and Gavin, M (2011), A desert in the delta: Participatory assessment of changing livelihoods induced by commercial shrimp farming in Southwest Bangladesh, in, Ocean & Coastal Management, Volume 54, Issue 1, January 2011, Pages 45-54. Available at: http://www.sciencedirect.com/science/article/pii/S0964569110001614#

[25] Alamgir, Fariba (2010), Contested Waters, Conflicting Livelihoods and Water Regimes in Bangladesh, unpublished research paper of MDS programme, ISS, The Hague, The Netherlands, November, 2010

[26] Solidarites International &Uttaran (2012), Chronic Poverty in the Southwest Coastal Belt of Bangladesh. Source:http://www.eldis.org/vfile/upload/1/document/1302/Chronic_Poverty_in_the_Southwest_of_Bangladesh_Article_11Dec12.pdf

[27] Rahman, islam 2018 M. A. Rahman and M. N. Islam (2018), Scarcity of Safe Drinking Water in the South-West Coastal Bangladesh, J. Environ. Sci. & Natural Resources, 11(1&2):17-25, 2018 ISSN 1999-7361 17

[28] Datta, D. K., Ghosh, P. K., Md. Rezaul Karim, Md. Mujibor Rahman (2009), Geochemical options for water security in a coastal urban agglomerate

of Lower Bengal Delta, Bangladesh, Journal of Geochemical Exploration, Elsevier. Journal homepage: www.elsevier.com/locate/gexplo

[29] Aggarwal, P.K., Basu, A.R., Poreda, R.J., Kulkarni, K.M., Froehlich, K., Tarafdar, S.A., Ali, M., Ahmed, N., Hussain, A., Rahman, M., Ahmed, S.R., 2000. A Report on Isotope Hydrology of Groundwater in Bangladesh: Implications for Characterization and Mitigation of Arsenic in Groundwater. (Report prepared within the IAEA-TC project BGD/8/016, Dhaka, Bangladesh).

[30] Shamsudduha, M., Taylor, R.G., Ahmed, K.M., Zahid, A., 2011. The impact of intensive groundwater abstraction on recharge to a shallow regional aquifer system: evidence from Bangladesh. Hydrogeol. J. 19, 901-916. https://doi.org/10.1007/s10040-011- 0723-4

[31] George, G.J., 2013. Characterization of Salinity Sources in Southwestern Bangladesh Evaluated through Surface Water and Groundwater Geochemical Analyses. Unpublished MSc Thesis. Faculty of the Graduate School, Vanderbilt University, USA

[32] Sultana, S., Ahmed, K.M., Mahtab-Ul-Alam, S.M., Hasan, M., Tuinhof, A., Ghosh, S.K., Rahman, M.S., Ravenscroft, P., Zheng, Y., 2015. Low-cost aquifer storage and recovery: implications for improving drinking water access for rural communities in coastal Bangladesh. J. Hydrol. Eng. 20, B5014007. https://doi.org/10.7916/D8280709.

[33] WaterAid (2017), Synthesis Report Climate resilient drinking water infrastructure based on a demand-supply and gap analysis For 39 Unions of 5 Upazilas under Khulna and Satkhira District of southwest coastal Bangladesh, an unpublished report, June 22, 2017

[34] MPO (Master Plan Organisation). 1987. Surface water availability. Technical report No. 10. MPO, Dhaka

[35] Joypasa (2009), Scarcity of drinking water; in perspective of Southwest Coastal region of Bangladesh. *Available at: http://community.eldis.org/joypasa/Blog/Scarcity-of-drinking-water--in-perspective-of-Southwest-Coastal-region-of-Bangladesh* Jul 13, 2009

[36] Nahian, M.A., Ahmed, A., Lázár, A.N., Hutton, C.W., Salehin, M. and Streatfield, P.K., 2018. Drinking water salinity associated health crisis in coastal Bangladesh. *Elem Sci Anth*, 6(1), p.2. DOI: http://doi.org/10.1525/elementa.143

[37] FAO (2015), Mapping Exercise on Waterlogging In South West of Bangladesh Mapping Exercise on Waterlogging In South West of Bangladesh, an unpublished document of Food And Agriculture Organization Of The United Nations, March 2015

[38] Nowreen, S., Jalal, M. R., and Khan, M. S. A (2014), Historical analysis of rationalizing South West coastal polders of Bangladesh. Water Policy 16 (2014) 264-279

[39] Tutu, Ashraf-Ul-Alam Tutu (2009), Tidal River Management in Bangladesh: People's Initiative on the Coastal River Basin Management to Solve Waterlogging in the Southwest Coastal Region of Bangladesh, In Water for the People: People's Water Resource Management Strategies, an unpublished document of Water for the People Network, IBON International, The Philippines.

Chapter 3

Natural Polonium-210 in Bivalve Species in Peninsular Malaysia Waters as Recent Pollution Indicator

Nurhanisah Zakri and Che Abd Rahim Mohamed

Abstract

Po-210 is an alpha rays emitter in U-238 decay series and a natural radionuclide found in the ocean, and bivalve is the best biological indicator compared to the other organisms because of their feeding methods that are filter-feeding and suspension-feeding. They are able to accumulate toxic substances from marine environment in their tissue and researches were conducted in edible tissues of *Meretrix meretrix, Perna virid, Glauconome virens, Anadara granosa, Anadara ovalis, Pholas orientalis, Donax* sp., *Polymesoda bengalensis, Phapia undulata,* and *Tellina virgate.* Result showed Po-210 activity distributions were ranging from 2.61 ± 1.50 to 517.46 ± 56.64 Bq/kg. The lowest value of Po-210 activity recorded in *Anadara granosa* and the highest value recorded in *Donax* sp. Small-sized of bivalve species contained higher Po-210 activity than the larger one. Higher Po-210 contents in bivalve obtained from the west coast of Peninsular Malaysia might be closely related to anthropogenic factors from the coastline. This study also found that *Donax* sp. is able to be a good indicator of environmental pollutants as it accumulates Po-210 in higher concentrations than other bivalve species. *Donax* sp. can be found in several parts of Malaysia and available in large quantities but it appears to be seasonal. While for seafood safety monitoring, *Anadara granosa* is capable of becoming a good benchmark for seafood security as it found in most parts of Malaysia. It is not seasonal and a kind of Malaysian favorite seafood.

Keywords: Polonium-210, bivalves, species, pollution, seafood

1. Introduction of Polonium-210

Polonium (Po) is a highly radioactive and semi-metal element that is rarely encountered with the symbol Po and atomic number 84. Polonium is a radioactive element that occurs naturally in very low concentrations in the earth's crust (about one per million trillion). Po has a low melting point and reactive metal in its pure form. Over 25 isotopes of polonium are known, with atomic masses between Po-192 to Po-218 (isotopes are different forms of an element that have the same number of protons in the nucleus but different numbers of neutrons). All isotopes of polonium radioactive but only three have a long half-life that enables to do research which is polonium-208 (42 years), polonium-209 (2.9 years), and Po-210 (138.4 days).

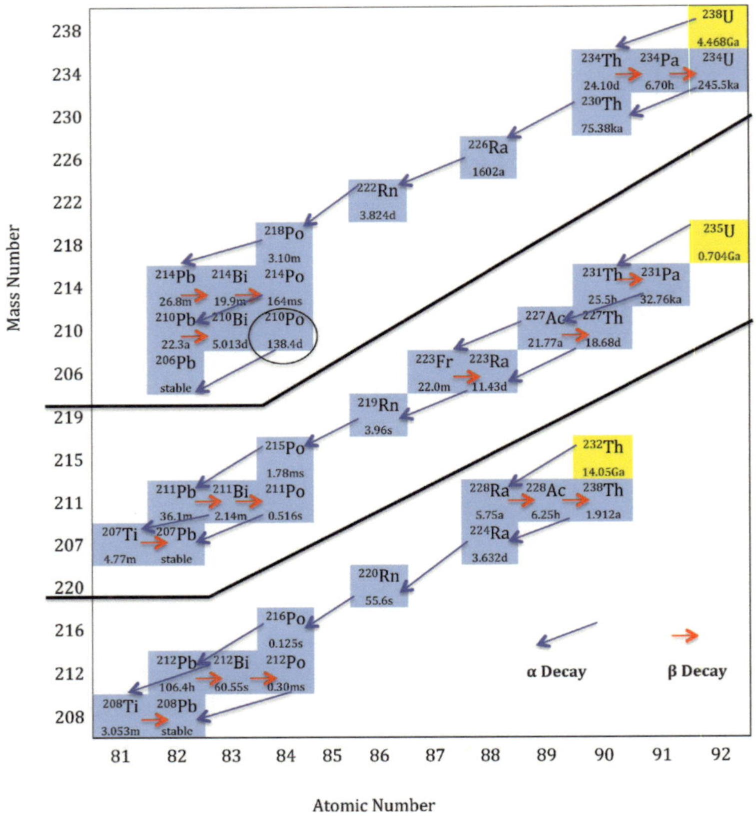

Note: The only long lived polonium isotope, ^{210}Po, is circled.

Figure 1.
The naturally occurring U-238, U-235 and Th-232 decay series, all of which contain polonium isotopes. Source: Carvalho [1].

Po-210 is a natural isotope that has a low melting point, volatile with 50% will vaporize at a temperature of 55° C within 45 hours. Po-210 is soluble in dilute acid but partially soluble in alkali. Po-210 is a product of the radioactive decay series of uranium-238 together with lead-210, it is one of the two products of decay of radon-222 and it decays to lead-206, which generate stable alpha particles. The energy released by the decay is more than 140 watts and temperature reach above 500° C.

Polonium-210 (Po-210) is produced by the natural decay of uranium-238, released widely in small amounts in the earth's crust (**Figure 1**). Uranium ores contain less than 0.1 mg of Po-210 per ton due to lack of Po-210 source. It will produce by synthetically in nuclear reactors by bombarding bismuth-209 (stable isotope) with neutrons. This form bismuth-210, which has a half-life of 5 days and beta emitter decays to produce Po-210. A large number of milligrams of Po-210 can be produced in addition to Po-209 and Po-208 by bombardment of bismuth-210.

Po-210 is usually used as a chemical to eliminate the effects of static electricity in the industrial paper processing machinery, sheet, plastics and synthetic fibers [2]. Po-210 is generally coated on a support foil and put in the brush holder, tube or other. The alpha particles from polonium ionize adjacent air and air ions then neutralize static electricity on the surface in contact with air. These devices typically need to be replaced every year because of the short half-life of this radioisotope. Po-210 is also used in brushes to remove dust from photographic film and camera lenses. Static electricity eraser materials usually contain tens to hundreds of Po-210 mCi. Po-210 can also be combined with beryllium to produce neutron sources and used as a catalyst for the production of neutrons of the first phase in the creation of atomic weapons [3]. Moreover, Po-210 is also a source of heat for the thermoelectric device.

In the environment, the contaminants of Po-210 is present as naturally and synthetically. Po-210 is highly toxic to human health because it has a high affinity and actively bind tightly and gathered on the particles and tissues of the organism addition to being the main contributor with 90% of the radiation dose and toxicity of naturally accepted by many marine organisms [1]. Humans are exposed to natural radiation in the high rate is through the consumption of food, especially seafood because marine organisms have a high capacity of bioaccumulation of radionuclides and other toxic elements from water. Determination of radioactivity in seafood such as crustaceans, mollusks and fish can be an indicator of the ecosystem and food security situation at the time. The level of this element influenced the quality of the marine environment, marine life and affect human health as the highest trophic level of the food chain. Then the objectives of this writing are (i) to determine the distribution of Po-210 in tissues of different species of bivalves, (ii) to compare the Po-210 concentrations in the tissues of bivalves from different locations, and (iii) to identify the pollution sources of Po-210 in the bivalves species.

2. Polonium-210 in marine ecosystem

2.1 Environment

In the vicinity, Po-210 is naturally present in very low concentrations. Po-210 is produced from the decay of radon-222, it can be found in the atmosphere. Po-210 is also released into the atmosphere during processing phosphate rock to elemental phosphorus. Previous studies have found the deposition from the atmosphere on tobacco leaves are in high concentrations [4]. The content of Po-210 in tobacco smoke is also high and the health impact is greater in smokers than non-smokers. In the marine environment, Po-210 is produced in large quantities by the decay of Pb-210 from the atmosphere. Then small amounts of Po-210 in the ocean are comes from the atmospheric Po-210. While, Po-210 concentrations in river and sea is usually widely varied as they involve the geological system and weathering processes of an area. The concentration of Po-210 is also increased by the presence of industrial activity around the river and the sea [5]. Polonium has existed widely in the environment and the natural background radiation wildlife. Therefore, the study and measurement concentration of polonium is important not only for physical health but also important in the field of geochronology and environmental science.

2.2 Water

Water covers 71% entire surface of the earth and an important medium ecosystem. All the natural elements dissolved in water and water into a source of nutrient storage and transfer medium of nutrients in the ecosystem abiotic and biotic [6]. Then, Po-210 has the same features as the nutrients contained in the surface waters where nutrients and Po-210 is taken as food by phytoplankton and released to other areas such as the euphotic zone [1]. Based on a study by Hong et al. [7] along Japanese waters, Po-210 concentrations in water columns decrease with depth and from winter to summer. Most studies Po-210 and Pb-210 found, the ratio of Po-210/Pb-210 in seawater is less than 1.0 and Po-210 becomes more reactive to organic particles than particles of inorganic [8, 9]. Po-210 activity against organic particles resembling the profile element where the concentrations of nutrients in surface water is low and increased in the mid-depth and reduced to a maximum depth [9, 10]. Po-210 concentrations in seawater are usually depending on the season, chemical and biological factors [7].

2.3 Sediment

Sediment is broken stone fragments that are chemically or by natural weathering. Sediment is the transition between the sandstone. Sediment deposition can occur anywhere and the sea is the last settlement. Sediment interaction with living organisms in aquatic ecosystems is high. There are nutrients in sediments and to be medium for marine life such as seaweed and benthic survive [11].

While sediment is one of the essential components for aquatic life and also a place for accumulate pollutants. Sediment can be an indicator of pollution due to contaminants in aquatic systems deposited on the seabed. Accumulation of pollutants in sediment at the seabed is important for pollution studies. Feature Po-210 has a high affinity bind to the particle that makes the sediment as the main source of contamination indicator Po-210 [12].

Generally, the content of Po-210 increased with increasing amount of silt, clay and organic matter [13]. Po-210 concentrations were high in the sediments occurs as sedimentary fragments of organic waste. Then, sediments rich with Po-210 is an important medium for removal polonium from water column into the organisms [14].

2.4 Marine flora

Marine flora is an important resource in the stability of marine ecosystems and phytoplankton is a primary source in the entire food chain of marine life. Phytoplankton accumulate Po-210 were found in the water column without involving light energy and temperature. This method involves passive adsorption on the surface of cells and cell control the intake of Po-210 in phytoplankton [15]. Po-210 concentrations in marine phytoplankton are varied and depend on the size and composition of the proteins in the cell [15]. Analysis by Cherry and Shannon [16] found the concentration range of Po-210 in phytoplankton at the Peru waters is ranging from 32 Bq/kg to 132 Bq/kg, and studied by Heyraud and Cherry [17] is about 237 Bq/kg. In addition, analysis by Folsom and Beasley [18] found concentration of Po-210 is too low, so the 0:07 Bq/kg in dinoflagellate taken from waters off California are experiencing an explosion of algae bloom.

Laboratory studies have found the accumulation of Po-210 by bacteria and phytoplankton depends on the cell structure protein [19, 20] and it is consistent with studies that have been conducted in animal protein [21, 22]. Seaweed or algae is rich in protein but Po-210 concentration is high in seaweed [14]. The level concentration of Po-210 is high in seaweed during winter and low during summer [23]. Finally the levels of Po-210 in marine biota were also influenced by the changing seasons with ranging from 10 Bq/kg to 70 Bq/kg especially in crab, squid and fish [17].

2.5 Edible seafood - bivalve

The feeding activities of bivalve usually proportion the contents of water column and sediment–water interface of seabed. These relationships provide ecosystem services that affect the entire food chain. Bivalves are an essential component in the river and ocean because it can act as a filter for bacteria, algae and other small particles. The ability of bivalves as a natural filter improves water quality, particularly marine ecosystems. Malaysian water is rich coastal area of sandy and muddy mangrove habitat is suitable for a variety of flora and fauna. Malaysia's tropical climate is hot and humid all year creating diversity on land and sea. This makes Malaysia is rich in marine life, especially species of shelled bivalves. Species of bivalves are often used as a food source population is as clams (*Meretrix meretrix*), Mussels (*Perna viridis* & *Glaucoma virens*), Shellfish (*Anadara granosa*), Conch shell (*Anadara ovalis*), Mentarang (*Pholas orientalis*), mussels (*Donax* sp.), seashell mangrove (*Polymesoda bengalensis*) and lala (*Phapia undulate* & *Tellina virgata*) and others can be found throughout Malaysia, especially coastal areas of Peninsular Malaysia (**Figure 2**).

Previous studies have found particularly bivalves *Mytilus* sp. able to accumulate toxic heavy metals found in the environment [24, 25]. These organisms accumulate many contaminants found in the water column as well as a bio-indicator contamination of an area [26]. Many studies have been conducted and found that alpha particles naturally released Po-210 accumulated in the tissues of various marine organisms, especially in organisms at high levels and concentrations higher than Pb-210 [22, 27] and tissue hepatopancreas in animals marine invertebrates has the highest concentration of natural dose in the marine ecosystem [17, 28, 29]. In the United Kingdom, researchers are focusing on fish and shellfish in areas that are not contaminated to trace the source of natural radionuclides produced by anthropogenic. The concentration of radionuclides Pb-210, Ra-226, U-238 and Th-232 in marine organisms is much lower than the concentration of Po-210 and varied according to diet of the users [30]. The study by Carvalho [31] found that the main exposure Shaheof Po-210 is through the consumption of seafood, the accumulation of these radionuclides is three times higher than people who consume foods mainland.

Bivalve is an organism that takes phytoplankton and suspended particles in the bottom of the sea as a food source. The variations levels of Po-210 in the edible tissue of bivalve is always different based on the digestive technique either using filtration or suspension method [32, 33]. Bivalves that do food intake filtration techniques have concentrations of Po-210 higher than the suspension food intake [34].

Connan et al. [23] found that the concentration of Po-210 in bivalve is highly during winter than summer season. This situation points out that the bivalves less accumulated Po-210 during summer than winter season. In winter season bivalves will produce the gametogenesis and reproductive processes will occur during the summer period [35]. Then there is the interconnection between Po-210 concentrations in bivalves with a variety of physiological, biochemical and sexual abuse that are reported to the concentration of heavy metals in bivalves [36]. Po-210 concentration is also dependent on the sampling location. The study conducted

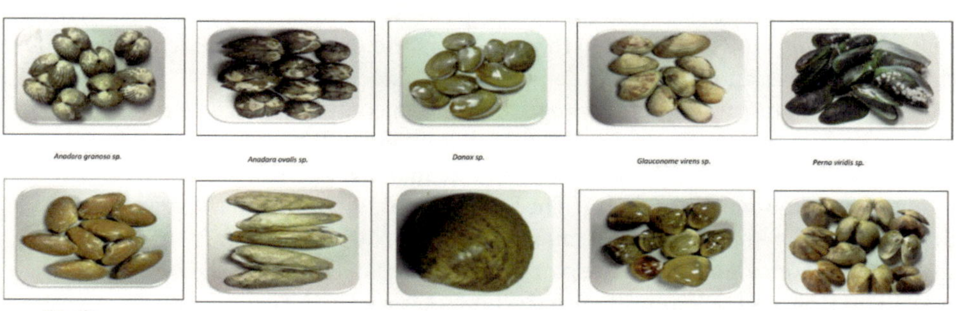

Figure 2.
Various edible species of bivalve in Malaysia waters.

by McDonald et al. [37] in six different sampling locations in Scotland, England, France and Monaco have found concentrations of Po-210 in the soft tissues of the organism is in the ranging from 111 Bq/kg to 459 Bq/kg and depending on the size of bivalve [38]. The level of Po-210 in edible tissue usually increased with decreasing size of bivalve as well reported by Bustamante et al. [34]. But the contrast findings reported by Conan et al. [23], where the larger size of bivalve containing high level Po-210. The difference of this study was more pronounced with age and associating the difference of each species of bivalves metabolism.

The distribution contents of Po-210 in tissue as well as organs with different species, life and habitat of bivalves are not well documented. According to Wildgust et al. [39], digestive organ tissue is about 10% of the total body weight but from that 15–36% is containing Po-210. The digestive system is the main route of Po-210 entering into bivalve body from marine environment. The highest level of Po-210 was recorded in the digestive tissues of *Chlamys Vaira* [34]. The high concentration of Po-210 has also been found in the digestive tissues of several species such as *M.trossulus, P.yessoensis* and *B.cornutus* [40, 41]. The level of Po-210 is different based on the function of organs as well described by Connan et al., [23].

In *Chlamys islandicus*, Po-210 concentrations were the highest recorded in the gill tissue, where the gill is the first organ of polonium entering into the bivalve body [34]. However Connan et al. [23] reported the digestive organ, gills and mantle of oyster containing highest level of Po-210 among other tissues. The concentration of Po-210 in soft tissue of *Chlamys varia* is two to three times higher than *Mytilus edulis* and suitable as pollution indicator but Po-210 concentrations recorded by this species is lower than *Chlamys varia* [34]. Po-210 concentrations in the digestive system was strong relationship with the changing contents of suspended particles, then proposed a major input Po-210 in the organism is of leachate sediment particles in the water column.

3. Human health and pollution indicator

3.1 Human health

Alpha particles are able to penetrate several sheets of paper or epidermis of the skin and can be stopped by clothing and skin. Thus, alpha radiation represents internal contamination and internal radiation hazard only if the particles inhaled, ingested or injected and penetrated through the opening cuts. By weight, Po-210 is 250,000 times more toxic than hydrogen cyanide and one gram of Po-210 can result in 50 million to 50 million of morbidity and mortality in humans [42]. Po-210 only becomes carcinogenic when introduced into the body. External physical exposure is less dangerous. The primary means of exposure is through ingestion of food and water as well as inhalation of air containing Po-210. Animal studies have found that 50–90% of Po-210 was taken through food and drink will soon leave the body through the feces and the remainder enters the bloodstream [43].

Generally, lung and kidney is important organ compare with others organ in human body. Almost 45% of Po-210, which is eaten, will be stored in the lungs, kidneys and liver, while 10% is stored in bone marrow and the remainder is distributed throughout the body [1]. Through breathing, Po-210 in the air will be deposited in the lungs. Po-210 is inhaled during breathing, either from radon in the air or cigarette smoke, can be stored in the mucous lining of the respiratory tract. The alpha particles emitted in the lungs can cause the cells lining the airways disturbed and damaged. Cell damage can potentially lead to lung cancer. The effect usually occurs in the kidney compared to the lungs, although higher doses in the lungs. Alpha particles are free from Po-210 can interrupt the system by destroying

DNA cells, modifying the structure and function of cells and cause mortality cells. Previous studies have found the risk of mortality from cancer can be occurred and Carpenter et al. [44] reported the radiation of these radionuclides would be affecting the tissue and genetic cause by the alpha particles emitted.

3.2 Po-210 as polluter and pollution indicator

Pollution is the entry of pollutants such as chemicals, noise, heat, light and energy into the environment resulting in destructive effect to endanger human health, natural resources and threatened ecosystems and interfere with the amenity and environmental use. Definition of pollution more informed in accordance with the Environmental Quality Act 1974 which states that pollution is any change either directly or indirectly to the physical properties, chemical, biological or radiation level of any part of the environment with the release, issue or placing waste to the detriment of beneficial uses, which gives rise to a dangerous situation or may be harmful to the health, safety or welfare or other organisms, plants and animals.

Water and ocean play an important role in controlling the balance of wildlife and the environment. Compared pollution on land can be seen clearly and more easily manage, pollution at sea cannot be delimited. Its negative impact will occur globally and are rapidly but cannot be seen with the naked eye. Sea was made barrels such as plastic waste, industrial waste and oil. This situation led to disruption of the ecosystem, destroying habitat and marine life. Even in small concentrations, pollutant toxic components capable of retarding the ability of marine life to breed and grow. The fact that the toxic substance decomposes difficult due to long life than through the food chain causing it to accumulate and poison the animals, especially marine shellfish and its impact will be seen in the highest levels of human food [45].

Radioactive contamination is usually expressed in units of radioactivity per unit area but for international unit (SI) is the Becquerel per square meter (Bq/m^2). The SI unit for measurement pollution in the organism is a unit of radioactivity per unit weight of organisms as Bq/kg. Radioactive contamination may be fixed or removable. In the case of fixed contamination, the radioactive material is distributed by definition, but still measurable. Monitoring involves the measurement of radioactive contamination or radiation dose of radionuclides associated with the assessment or control of exposure to radiation or radioactive substances and the interpretation of results [46]. Methodological and technical details for the design and operation of environmental radiation monitoring program at different types of radionuclides will be guided by the International Atomic Energy Agency standard protocol.

As reported by Utusan Malaysia [47], industrial waste pollution flowing into the sea in Malaysia, especially in the industrial states of Penang, Selangor and Johor had to be addressed. This is because the pollution can affect marine life and the next source of food and traditional fishing economy. For example, the local shellfish contain heavy metals lead and other toxic substances to be used as case studies of high institutes of higher learning and overseas. In fact, many may recall, not long ago there were countries that had prevented the importation of scallops from our country. Another consequence, shellfish breeders Sungai Juru, Penang, which is a major producer of oyster country, reported losses of up to half of the shell as a result of death, death or disability due to the quality of seawater in the state is too bad. In fact, the actual farming shellfish aquaculture is one of the main branches of the state, with cockles *Anadara* sp. 40,000 tonnes in 1991 [48].

The seafood and the result is the main source of protein coastal population. Toxic pollutants and radionuclides are present in marine environments can also exist and be detected in the tissues of marine life. The concentration of toxic substances in the tissues of marine life increases with increasing trophic level. Po-210 radioisotope study in bivalves has been widely carried out abroad. In Malaysia, the research focused on the determination of trace elements, metal and heavy metal and radionuclide studies to be lacking. The main natural radiation exposure is through the consumption of seafood containing radioactive elements [49]. Therefore, many countries and international organizations have been monitoring and determining the health risks of seafood through diet by population and dose seafood safety by the use of human [30, 40, 50–52].

Today many researchers conducting a study on natural radionuclides e.g., Po-210 because a lot negative impact on human health were reported. Previous studies have found marine organisms accumulate Po-210 in high concentrations and food chain is one of the main routes of distribution and accumulation of Po-210 in marine organisms (e.g. [21, 53, 54]).

Ongoing studies towards the marine environmental conditions are important to all countries in an effort to reduce and prevent pollution from becoming widespread. Therefore, constant and systematic monitoring is necessary. The usage of marine organism as a bio-indicator for heavy metal pollutions has shown promising results [8]. Bivalve of *Mytilus* sp. has been a good pollution indicator and is widely used due to its ability to accumulate heavy metals from its environmental habitat [24].

Recently, biological indicator has been used to monitor the concentration level of heavy metals and both the stable and unstable radioactive materials in the marine environments in which has become the norm for researchers worldwide. Organisms capable of accumulating pollutants such as bivalve are preferred as indicators in order to determine the presence of specific pollutants in the environment.

The main characteristics for the bio-indicating organisms are the ability to accumulate widely-spread pollutants, can be easily found throughout the region geographically, may survived a whole year round, highly sensitive and tolerant of pollutants without affecting the organisms itself, may be easily obtain for sampling and preservation as well as showing positive correlation between pollutant concentration and environmental pollution [55]. In addition to its ability to act as a pollutant source and environmental quality bio-indicator, it also allows comparisons between elements such as heavy metals and radioisotopes in organisms from different geographical areas.

4. Sampling and methodology

Malaysia has nearly 4800 km of coastline area that includes sandy beaches and muddy beaches. The region is rich in rich resources and livelihoods that are vital to the stability of the ecosystem and contribute to the national economy. This makes Malaysia rich in marine life, especially for bivalve species. These species can be found throughout Malaysia, especially the coastal areas of Peninsular Malaysia.

The sampling has been carried out carried out during March to December of 2012 around the shoreline areas of interest around Peninsular Malaysia and the sampling location is as shown in **Figure 3**. The bivalve samples were purchased from around the coastal areas of each state (**Table 1**).

The samples obtained were frozen and taken back to the Chemical Oceanography Laboratory, Faculty of Science and Technology, Universiti Kebangsaan Malaysia, Bangi, Selangor. In the laboratory, the species of each sample was identified. A total of 10 species per species were taken at random for analysis. Once the species was identified, the samples were cleaned from sediment, epiphyte, epifauna and washed with distilled water. The soft tissue portion of the bivalve was removed and separated from the shell. The weight of wet tissue was measured using an electronic

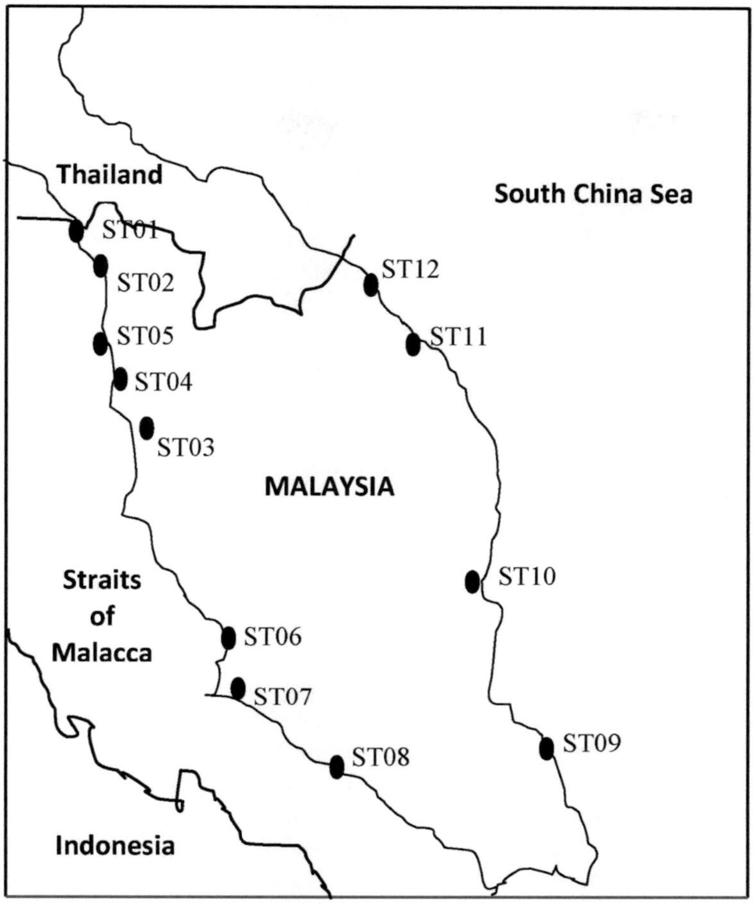

Figure 3.
Sampling location conducting during this study.

State	Station	Location	Date of sampling
Perlis	ST01	Penjaja Kuala Perlis	April 2012
Kedah	ST02	Pasar Basah Kuala Kedah	April 2012
Perak	ST03	Pasar Kuala Gula	April 2012
Pulau Pinang	ST04	Pasar Basah Juru, Seberang Perai	April 2012
	ST05	Pasar Bayan Baru	July 2012
Selangor	ST06	Penjaja Pantai Remis, Kuala Selangor	September 2012
	ST07	Pasar Basah Tanjung Karang	September 2012
Melaka	ST08	Penjaja Sebatu	December 2012
Johor	ST09	Pasar Basah Mersing	March 2012
Pahang	ST10	Pasar Nelayan Berserah, Kuantan	July 2012
Terengganu	ST11	Penjaja Setiu, Kuala Terengganu	October 2012
Kelantan	ST12	Pasar Siti Katijah, Tok Bali	July 2012

Table 1.
List of sampling sites were conducted during this study.

scale and readings were taken. Then, the samples were dried in an oven at 60°C for 24 hours to obtain the dry weight of the tissue. Once dried, the dry weight of tissue is taken. The difference in weight is between 15 and 20%. The dried samples were incised using mortar and stored in aluminum foil for radiochemical analysis.

A modified radiochemical separation method has been used for Po-210 analysis in the organism samples (e.g., [56–58]). The known Po-209 traces were added to 0.5 g of dried samples. Then, the samples were dissolved and digested using HNO_3 and H_2O_2. Once digested, the solution is filtered and heated to a moderate temperature for the evaporation process to occur till it dries. The dried sample solution was dissolved in 50 ml of 0.5 M HCl. The ascorbic acid was added to lower Fe (III) and Po-210 was deposited on silver discs of 2 cm in diameter for 3 to 4 hours at 70 to 90° C. The silver discs were then dried and the Po-210 activity was calculated with the Alpha Spectrometer system. Po-210 radiochemical analysis was performed within 2 months from the date of sampling and activity was calculated on the sampling date. The quality of the method and analysis were determined using the IAEA-134 standard reference material.

5. Distribution of Po-210 in the edible tissues of bivalve species from Malaysian waters

The results showed that the concentration range of Po-210 in the studied bivalves was in between of 2.61–517.46 Bq/kg based on **Table 2**. The lowest Po-210 concentrations in bivalve tissue were recorded in *Anadara granosa*, and the highest values were recorded in *Donax* sp. In general, it can be observed that the concentration of Po-210 in the bivalve tissue decreased with increasing tissue weight and shell size. This relationship is shown in species *Anadara granosa, Anadara ovalis, Meretrix meretrix, Paphia undalata, Pholas orientalis* and *Polymesoda bengalensis.* The concentration of Po-210 was found to be higher for the smaller-sized individuals compared to larger ones. Based on Ryan et al. [38], the Po-210 concentration depends on size. The results are concordant with previous study [34, 46, 59].

Bivalves filtering water from environment for food intake which Po-210 present in the environment [60]. The concentration of Po-210 in bivalves depends on the content of Po-210 found in food [61], the rate of food intake, the degree of absorption of food by bivalves [62] and the rate of Po-210 metabolism [46]; all these

Species	Size range (mm)	Range of Po-210 concentration (Bq/kg)
Anadara granosa	15–32	2.61–308.54
Donax sp.	15–24	34.34–517. 46
Anadara ovalis	21–56	3.03–221.65
Meretrix meretrix	25–57	12.45–419.12
Perna viridis	55–80	12.45–41.10
Paphia undulata	30–45	28.0–288.0
Pholas orientalis	65–104	18.13–200.2
Glauconome virens	22–35	47.71–234.97
Polymesoda bengalensis	55–66	4.92–50.14
Tellina virgata	29–39	31.37–169.54

Table 2.
The range of size and Po-210 concentration for ten species in this study.

factors depend on the size and weight of bivalves. The concentration of Po-210 in small-sized bivalve tissues is higher because it has a higher rate of metabolism to grow and its nutritional activity is higher than that of larger and older bivalves [63]. In terms of general physiology of life, the small size of organism is still in juvenile stage, therefore, has yet to have a matured organ system. The juvenile organ system is unable to function properly as it contains substances that cannot be absorbed by the juvenile organ and accumulates in the body [46].

Moreover, physically smaller size and weight contributes to a smaller volume. When the amount of Po-210 read as per the volume of bivalve will give a greater concentration reading value then a higher concentration of Po-210 is found in the bivalve smaller in size with low volume. A factor that also contributes to the high concentration of Po-210 in small-sized bivalve species is movement. The smaller and lighter size facilitates the movement of bivalve, the more movements performed the wider the area traversed then the more Po-210 from the environment will accumulate in the tissues.

However, it differs with *Donax* sp., *Perna viridis, Pholas orientalis* and *Glauconome virens* in which has shown that an increased size of the shell exhibits higher concentration of Po-210 in the tissue. As for *Pholas orientalis* and *Tellina virgate,* the level of Po-210 is proportion with tissue weight as similar as published by Conan et al. [23]. According to the study conducted, larger-sized bivalves have higher concentrations of Po-210. The differences in the results of our study are more clear shows by relating the effects of age and the difference in metabolic rate of each species of bivalve.

In addition, the physical characteristics of the thickness and physical structure of the shell may also be contributing [64]. Notably, *Donax sp.* and *Pholas orientalis* has thin and consistent shells for every size. Physically, the absorption of Po-210 from the environment is high at a larger surface area. *Donax* sp. and *Pholas orientalis* the larger ones have a surface area for greater absorption. Thus, the concentration of Po-210 is higher with increasing size. The feature of having a consistent shell thickness for each size can also be observed in *Perna viridis* However, these features are not seen in *Tellina virgata* and *Glauconome virens* because it has a thick shell and its thickness increases with increasing size.

5.1 Concentration of Po-210 in different species of bivalve tissues

Based on **Figure 4**, it can be seen that the highest average Po-210 concentration in the bivalve tissues was recorded in *Donax* sp., followed by *Meretrix meretrix, Glauconome virens, Paphia undulata, Tellina virgata, Anadara granosa* and *Pholas orientalis* with a Po-210 concentration more than 50.0 Bq/kg. On the other hand, the *Anadara ovalis, Perna viridis,* and *Polymesoda bengalensis* recorded an average less than 50.0 Bq/kg. The same characteristics can be observed in the five species that record the highest Po-210 concentration values are taxonomically from the order Veneroida. Veneroids generally have thick and muscular valve muscles of the same size [65], most of these species move actively against the sessile. It tends to be filtered eater and eat through chambers paired with suspension with folded gill structure features. Veneroida habitat is at the base of shallow substrate shallow water and sandy beaches [66]. For *Anadara granosa*, nutrient uptake from the water column and direct interaction with sediment on the seabed is a factor of high Po-210 concentration in its tissues [67]. Bivalve is an organism that takes phytoplankton and suspended particles on the seabed as food sources. The activity of Po-210 in organic particles resembles the profile of nutrient elements where the concentration is low on the surface of the water and increases at mid-depth and decreases at maximum depth [9, 10]. The variation rate of Po-210 concentration in

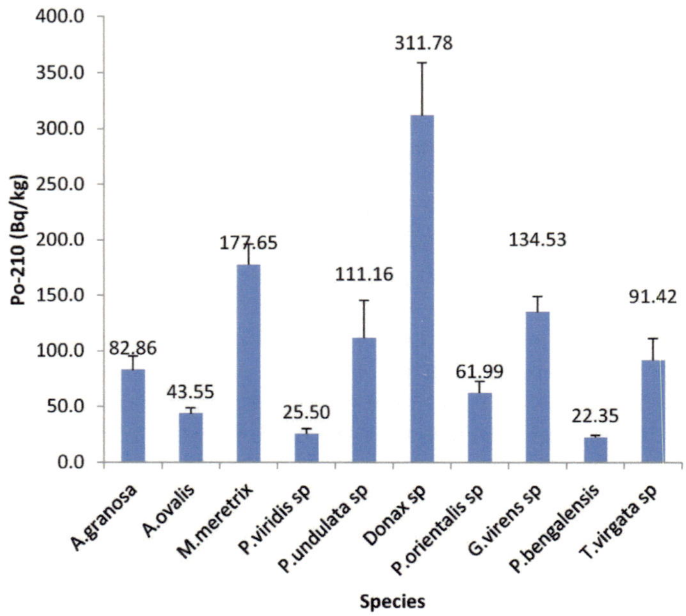

Figure 4.
Contents of Po-210 in edible tissue of bivalve species.

most bivalves is also high based on its nutritional techniques. Bivalve that performs filtered food intake technique has a higher content of Po-210 concentration than those that perform food intake technique by suspension [34].

The habitat of bivalves also influences the concentration of Po-210 in its tissue [67]. The results of this study found that bivalves living in muds such as *Pholas orientalis* and *Polymesoda bengalensis* have lower average readings compared to *Polymesoda bengalensis* that lives in mangrove swamps. In general, the content of Po-210 increases with increasing silt, clay and organic matter [13]. The high Po-210 concentration in the sediment is due to the sedimentation of organic debris. Therefore, sediment rich in Po-210 is an important medium for the transfer of polonium from water column to the organisms [14].

Differences in sizes may also contribute to the content of Po-210 concentrations in bivalve tissues. Small-sized bivalve species contain higher concentrations of Po-210 than larger ones. The average sample size of *Donax* sp., is 20 mm, followed by *Anadara granosa* with an average size of 25 mm, *Glauconome virens* with an average size of 29 mm, *Tellina virgate* with an average size of 33 mm and *Meretrix meretrix* with an average size of 34 mm. Meanwhile, the largest average size is the *Pholas orientalis* with an average size of 89 mm following *Perna viridis* with an average size of 62 mm and *Polymesoda bengalensis* with an average size of 60 mm. The same findings were obtained as distribution in each species, which the smaller the bivalve's size, the higher the Po-210 concentration in the tissue. The physical factors such as smaller size and weight also contributes to the concentrations. The small size gives the lower volume value, therefore the higher Po-210 concentration value will be found in the smaller sized bivalves tissue as the amount of Po-210 concentration obtained will be divided by the weight of the sample of the studied organism.

5.2 Contents of Po-210 at different pollution sources

The highest average concentration of Po-210 in the tissue of *Anadara Granosa* was recorded in samples taken from Kedah following by Johor and Perlis with the concentrations of >60.0 Bq /kg (**Table 3**). These three areas are located in the west coast of Peninsular Malaysia. The highest average concentration of Po-210 recorded in the tissues of *Anadara ovalis* from Melaka with an average concentration of 107 Bq/kg followed by samples from Perak at 41 Bq/kg. Whereas, for *Meretrix meretrix* the highest concentration was from the location of Kelantan, followed by samples from Perlis at a value more than 100 Bq/kg. For *Paphia undulata,* samples from the Pahang are deemed higher than the samples taken in Selangor. Lastly, a higher concentration was recorded in Perlis for *Pholas orientals* rather than samples from Selangor. Overall, the relationship between Po-210 concentration and sampling location is unclear as not all bivalves species are obtained from each location. However, the sampling location located west coast of Peninsular Malaysia recorded higher Po-210 concentration in most bivalves species studied. The concentration of Po-210 in seawater depends on the season, chemical and biological factors [68, 69]. Connan et al. [23] found that Po-210 concentration was higher during winter and lower in summer. This situation indicates that bivalves accumulate less Po-210 during the summer than in winter. Winter is the season for bivalves to undergo gametogenesis while reproduction process occurs throughout the summer [35]. Thus, there is an association between Po-210 concentration levels in bivalves with physiological, biochemical and sexual variations as well as seasonal changes. The discrepancy in the results of this study may be due to the sampling conducted throughout March to December 2012. The inconsistent weather changes throughout the year may be the factor for different Po-210 concentrations for each species from different locations. Previous study by Khan [46] showed Po-210 concentration among mussels varied significantly based on the season and the concentration is lesser during monsoon due to physiological and metabolic changes.

Besides that, the concentration of Po-210 also depends on the sampling location. A study conducted by McDonald et al. [37], six different sampling sites in Scotland, England, France and Monaco has shown that the concentration of Po-210 in the soft tissue of the organism was in the range 111–459 Bq/kg. The level of industrial development in an area contributes to the anthropogenic entry of Po-210 into the sea [1]. In this study, the Po-210 concentration values were higher in bivalves sample taken from west coast of Peninsular Malaysia rather than samples taken from the east coast location. The Straits of Malacca is a strategic location for major international shipping lane and the concentration of agriculture, industry and urbanization on the west coast of Peninsular Malaysia, which lead to the pollution [70]. The removal and release of toxic materials from ships may be a factor of high Po-210 concentration. The industries such as coal power station and processing factories at the west coast of Peninsular Malaysia discharge or release their effluents through air, river or sea and finally settle into the ocean. The effluent and radionuclide derivatives also contribute to the high concentration of Po-210 in bivalve's tissue. In general, samples obtained from west coast such as Perlis, Penang, Johor and Kedah gave higher readings of almost all bivalves species studied. No obvious factor can be discussed due to the lack of industrial area in Perlis. The possibility is probably due to both location geographically are close to Thailand. Seasonal factors and wind speeds may be associated with high Po-210 concentration content in samples from this location [71]. The transfer of particles containing Po-210 from the air, land or nearby ocean may occur because Po-210 is known to have a high affinity bound to the particles [12]. According to the study of Theng et al. [72], the concentration of Po-210 in clams in Kuala Selangor is based on environmental factors and sampling location. The content of the Po-210 concentration in the bivalve studied was different at different locations (**Table 4**). Mustafha et al. [80] stated the Po-210 concentration vary greatly in different locations.

State	Species of bivalve (Bq/kg)									
	A.granosa	*A.ovalis*	*M.meretrix*	*P.viridis*	*P.undulata*	*Donax* **sp.**	*P.orientalis*	*G.virens*	*P.bengalensis*	*T.virgata*
Perlis	98.09	24.21	187.63	—	—	—	71.66	—	—	91.42
Kedah	124.77	—	97.38	—	—	—	—	—	—	—
Perak	72.78	41.05	—	—	—	—	—	—	—	—
Pulau Pinang	93.53	—	—	—	—	—	—	134.53	—	—
Selangor	70.44	23.90	163.31	—	40.49	311.78	52.31	—	—	—
Melaka	16.48	107.16	—	—	—	—	—	—	—	—
Johor	105.64	—	—	25.50	—	—	—	—	—	—
Pahang	—	—	—	—	181.61	—	—	—	—	—
Terengganu	—	—	—	—	—	—	—	—	22.35	—
Kelantan	—	—	284.39	—	—	—	—	—	—	—

Notice: (−) samples unavailable during sampling.

Table 3.
Average contents of Po-210 in various species of bivalve from different states in peninsular Malaysia.

Location	Po-210 (Bq/kg)	References
Peninsular Malaysia	2.61–518	This study
Kapar, Malaysia	4.61–240	Alam et al. [71]
Cuba	21–30	Alonso-Hernandez et al. [56]
England	16–36	Young et al. [73]
India	305–597	Suriyanarayanan et al. [14]
Ribble Estuary	29	Rollo et al. [74]
Portuguese Coast	5.8–132	Carvalho [75]
Kuala Selangor, Malaysia	31.2–92.4	Theng and Mohamed [76]
Kudankulam Coast, India	5.4–248	Khan and Wesley [77]
Tiruchirappalli, India	57.42–106	Shaheed et al. [78]
Kalpakkam, India	35.19	Iyengar et al. [79]

Table 4.
Concentration of Po-210 in mollusks tissue results from studies around the world.

6. Conclusion

The study found that the concentration of Po-210 is varied in different species of bivalve. Biological, chemical and physical factors as well as seasonal and sampling location contribute to the different concentration of Po-210 in the bivalve tissues studied. The study found that the distribution range of the concentration of Po-210 in bivalves were between 2.61–517.46 Bq/kg. The lowest concentration of Po-210 in bivalve tissues recorded in *Anadara granosa* and the highest was recorded in *Donax* sp., Smaller size bivalve species contain high concentration of Po-210 compare to bivalve with larger size. In addition, high concentration of Po-210 in bivalve tissues from the west coast of Peninsular Malaysia were observed to have a relationship with the anthropogenic factors which is industrial activity and shipping routes along the coastal.

Recent finding in this research showed that for the observation of environment pollution specifically marine pollution *Donax* sp. can be a good indicator because it accumulated more Po-210 compared other species. Although, *Donax* sp. is present seasonally, it can easily found in Malaysia and large quantity. As for seafood safety monitoring, *Anadara granosa* able to act as a good indicator as it can easily found in parts of Malaysia all year and in addition to being the main food in Malaysia.

7. Suggestion for the future research

1. Further studies on the impact of taking seafood containing Po-210 can be carried out using the data from this research and based on the total daily intake of seafood by the locals. Through risk assessment, safety dose and the risk of disease can be determined.

2. More study should be done on *Donax* sp. including life cycles, contents of others trace elements and may be used as a good indicator of pollution.

3. Radionuclide monitoring studies need to be done in each species of marine organisms that act as the main seafood source for the local people.

Acknowledgements

The authors wish to thank all laboratory members of the Faculty of Science and Technology, Universiti Kebangsaan Malaysia for their help during this study. Thanks also to the Ministry of Science, Technology and Innovation for the financial support through grant code: 04-01-02-SF0801.

Conflict of interest

The authors declare no conflict of interest.

Notes/thanks/other declarations

Thanks.

Author details

Nurhanisah Zakri and Che Abd Rahim Mohamed*
Faculty of Science and Technology, Universiti Kebangsaan Malaysia, Bangi, Selangor, Malaysia

*Address all correspondence to: carmohd@ukm.edu.my

References

[1] Carvalho, F., Fernandes, S., Fesenko, S., Holm, E., Howard, B., Martin, P., Phaneuf, M., Porcelli, D., Prohl, G. and Twinning, J. The environmental of polonium. International Atomic Energy Agency, 2017; 484.

[2] Johansson, L.Y. Determination of Pb-210 and Po-210 in aqueous environmental samples. Doctoral dissertation, Hannover: Gottfried Wilhelm Leibniz Universität Hannover, 2008.

[3] Rhodes, R. The making of the atomic bomb. Simon & Schuster, London, 1986.

[4] Papastefanou, C. Radioactivity of tobacco leaves and radiation dose induced from smoking. International Journal of Environmental Research and Public Health, 2009; **6**(2): 558-567.

[5] Aysun Uğur, Güngör Yener, Asiye Basarı. Trace metals and Po-210 (Pb-210) concentrations in mussels (*Mytilus galloprovincialis*) consumed at western Anatolia. Applied Radiation and Isotopes, 2002; **57**(4), 565-571.

[6] Zaharescu, D.G., Burghelea, C.I., Dontsova, K., Presler, J.K., Hunt, E.A., Domanik, K.J., Amistadi, M.K., Sandhaus, S., Munoz, E.N., Gaddis, E.E. and Galey, M. Ecosystem-bedrock interaction changes nutrient compartmentalization during early oxidative weathering. Scientific reports, 2019; **9**(1): 1-16.

[7] Hong G.H., Kim Y.I., Baskaran M., Kim S.H., Chung C.S. Distribution of Po-210 & export of organic carbon from the euphotic zone in the Southwestern East Sea (Sea of Japan). Journal of Oceanography, 2008; **64**: 277-292.

[8] Kharkar D.P., Thomson J., Turekian K.K.,Forster W.O. Uranium & thorium decay series nuclides in plankton form the Caribbean Limnology & Oceanography, 1976**; 21**: 294-299.

[9] Ritchie G.D., Shimmield G.B. The use of the Po-210/Pb-210 disequilibria in the study of the fate of marine particulate matter. 1991; In P.J. Kershaw & D.S. Woodhead (eds.).

[10] Bacon M.P., Spencer D.W., Brewer P.G. Pb-210/Ra-226 & Po-210/Pb-210 disequilibria in seawater & suspended particular matter. Earth & Planetary Science Letters, 1976; **32:** 227-296.

[11] Dixit S.S., Witcomb D. Heavy metal burden in water, substrate, & macroinvertebrate body issue of a polluted river Irwell (England). Environment Pollution Series B, Chemical & Physical, 1983; **6**(3): 161-172.

[12] Shaheed, K., Somasundaram, S.S.N., Shahul Hameed, P. and Iyengar, M.A.R. A study of polonium-210 distribution aspects in the riverine ecosystem of Kaveri, Tiruchirappalli, India, Environmental Pollution, 1997; **95**: 371-377.

[13] Narayana Y., Rajashekara K.M. Study of Po-210 & Pb-210 in the riverine environment of coastal Karnataka. Journal of Environment Radioactivity, 2010; **101**(6): 468-471.

[14] Suriyanarayanan S., Brahman&han G.M., Malathi J., Kumar S.R., Masilamani V., Hameed P.S. Studies on distribution of Po-210 & Pb-210 in the ecosystem of Point Calimere Coast (Palk Strait), India. Journal of Environment Radioactivity, 2008; **99**(4): 766-771.

[15] Stewart G.M., Fisher N.S. Experimental studies on the accumulation of polonium-210 by marine phytoplankton. Limnology & Oceanography, 2003; **48**(4): 1720-1720.

[16] Cherry R.D., Shannon L.V. The alpha radioactivity of marine organisms. Atomic Energy Review, 1974; **12**: 3-45.

[17] Heyraud M., Cherry R.D. Polonium-210 & lead-210 in marine food chains. Marine Biology, 1979; **52**(3): 227-236.

[18] Folsom T.R., Beasley T.M. Contribution from the alpha emitter Po-210 to the natural radiation environment of the marine organisms. 1973; Proceedings of International Symposium on Radioactive Contamination of the Marine Environment, Seattle. Vienna : IAEA.

[19] Fisher N.S., Burns K.A., Cherry R.D., Heyraud M. Accumulation & cellular distribution of 241Am, Po-210 & Pb-210 in two marine algae. Marine Ecology Progress Series, 1983; **11**:233-237.

[20] Cherrier J., Burnett W.C., Larock P.A. Uptake of polonium & sulfur by bacteria. Geomicrobiology Journal, 1995; **13**: 103-115.

[21] Heyraud M., Fowler S.W., Beasley T.M., Cherry R.D. Polonium-210 in euphausids: A detailed study. Marine Biology, 1976; **34**(2): 127-136.

[22] Calvolha F.P., Fowler S.W. A double-tracer technique to the determine the relative importance of water & food as sources of polonium-210 to marine prawns & fish. Marine Ecology Progress Series, 1994; **103**:251-264.

[23] Connan O., Germain P., Solier L., Gouret G. Variations of Po-210 & Pb-210 in various marine organisms from Western English Channel: contribution of Po-210 to the radiation dose. Journal of Environment Radioactivity, 2007; **97** (2-3): 168-188.

[24] Boisson F., Cotret O., Fowler S.W. Bioaccumulation & retention of lead in the mussel *Mytilus galloprovincialis* following uptake from seawater. The Science of the Total Environment, 1998; **222**(1-2): 55-61.

[25] Da Ros L., Nasci C., Marigomez I., Soto M. Biomarkers & trace metals in the digestive gl& of indigenous & transplanted mussel, *Mytilus galloprovincialis*, in Venice Lagoon, Italy. Marine Environment Research, 2000; **50** (1-5): 417-423.

[26] Elliotta, P., Aldridgea, D.C., Moggridgeb, G.D. Zebra mussel filtration and its potential uses in industrial water treatment. Water Research, 2008; **42**: 1664-1674.

[27] Stepnowski P., Skwarzec B. A comparison of Po-210 accumulation in molluscs from southern Baltic, the coast of Spitsbergen & Saselki Wielki Lake in Poland. Journal of Environment Radioactivity, 2000; **49**: 201-208.

[28] Folsom T.R., Wong K.M., Hodge V.F. Some extreme accumulations of natural polonium radioactivity observed in certain oceanic organisms. In J. A. S. Adams, W. M. Lowder & T. F. Gesell (eds.). The Natural radiation environment, 1972; **11**: 863-882. Houston, NTIS: Rice University & the University of Texas.

[29] Cherry R.D., Heyraud M. Evidence of high natural radiation doses in certain mid-water oceanic organisms. Science, 1982; **218**: 54-56.

[30] Yu K.N., Moa S.Y., Young E.C.M., Stokes M.J. A study of radioactivities in six types of fish consumed in Hong Kong. Applied Radiation & Isotopes, 1997; **48**(4): 515-519.

[31] Carvolha F.P. Po-210 & Pb-210 in sediment & suspended matter in the Tagus estuary, Portugal. Local enhancement of natural levels by wastes from phosphate ore processing industry.

Science of the Total Environment, 1995; **159**(2-3): 201-214.

[32] Boyle P.R. Molluscs & Man. 1981; 134. Edward Arnold. London: The institute of Biology's studies in biology.

[33] McLusky D.S. The estuarine ecosystem-(Tertiaty level biology), 1981; Glasgow, UK: Blackie & Son Ltd.

[34] Bustamante P., Germain P., Leclerc G., Miram P. Concentration & distribution of Po-210 in the tissues of the scallop *Chlamys varia* & the mussel *Mytilus edulis* from the coast of Charente-Maritime (France). Marine Pollution Bulletin, 2002; **44**(10): 997-1002.

[35] Seed R. The ecology of *Mytilus edulis* L. (Lamellibranchiata) on exposed rocky shores, breeding & settlement. Oecologia, 1969; **3**: 277-316.

[36] Cossa D. A review of the of Mytilus sp. As quantitative indicators of cadmium & mercury contamination in coastal waters. Oceanologica Acta, 1989; **12**(4): 417-432.

[37] McDonald P., Jackson D., Leonard R.P., McKay K. An assessment of Pb-210 and Po-210 in terrestrial foodstuffs from regions of England and Wales. Journal of Environmental Radioactivity, 1999; **43** (1), 15-29.

[38] Ryan, T.P., Dowdall, A.M., McGarry, A., Pollard, D., Cunningham, J.D. Po-210 in *Mytilus edulis* in Irish marine environment. Journal of Environmental Radioactivity 1999; **43**: 325-342.

[39] Wildgust M.A., McDonald P., White K.N. Temporal changes of Po210 in temperate coastal waters. The Science of the Total Environment, 1998; **214**(1-3): 1-10.

[40] Yamamoto M., Abe T., Kuwabara J., Komura K., Ueno K., Takizawa Y. Polonium-210 & lead-210 in marine organisms. Intake levels for Japanese. Journal of Radioanalytical & Nuclear Chemistry, 1994; **178**(1): 81-90.

[41] Stepnowski P. Skwarzec B. Tissue & subcellular distributions of Po-210 in the crustacean *Saduria entomon* inhabiting the southern Baltic Sea. Journal of Environment Radioactivity, 2000; **49**(2): 195-199.

[42] Seiler, R. Po-210 in drinking water, its potential health effects, and inadequanvy of the gross alpha activity MCL. Science Total Environment, 2016; **568**: 1010-1017.

[43] Ansoborlo, E., Berard, P., Den Auwer, C., Leggett, R., Menetrier, F., Younes, A., and Moisy, P. Review of chemical and radiotoxicological properties of Polonium for internal contamination purposes. Chemical Research in Toxicology, 2012; **25**(8): 1551-1564.

[44] Carpenter L.M., Higgins C.D., Douglas J., Maconochie N.E., Omar R.Z., Fraser P., Beral V., Smith P.G. Cancer mortality in relation to monitoring for radionuclide exposure in three UK nuclear industry workforces. Br J Cancer, 1998; **78**(9): 1224-1232.

[45] Vilarino. N., Louzao, M.C., Abal, P., Cagidae, E., Carrera, C., Vieytes, M.R. and Botana, L.M. Human poisoning from marine toxins: unknowns for optimal consumer consumer protection. Toxins, 2018; **10**(8): 324.

[46] Khan, M.F., Wesley, S.G and Rajan, M.P. Polonium-210 in marine mussels (bivalve molluscs) inhabiting the southern coast of India. Journal of environmental radioactivity, 2014; **138**: 410-416.

[47] Utusan Malaysia. Utusan Malaysia article (8th Jun 2008), 2008; Website: www.utusan.com.my

[48] Hanafi, H.H, Arshad, M.A and Yahaya, S. Report on a regional study and workshop on the environmental assessment and management of aquaculture development (TCP/RAS/2253). NACA Environment and Aquaculture Development Series, 1995: **1**

[49] Mishra S., Bhalke S., Pit G.G., Puranik V.D. Estimation of Po-210 & its risk to human beings due to consumption of marine species at Mumbai, India. Chemosphere, 2009; **76**(3): 402-406.

[50] Smith J., Towler P.H. Polonium-210 in cartilaginous fishes (Chondrichthyes) from southern Australia waters. Australian Journal of Marine and Freshwater Research, 1993; **44**: 727-733.

[51] IAEA-TECDOC-838. Sources of Radioactivity in the marine environment & their Relative Contributions to Overall Dose Assessment from Marine Radioactivity (MARDOS), 1995; Vienna: IAEA.

[52] Al-Masri M.S., Mamish S., Budeir Y., Nashwati, A. Po-210 & Pb-210 concentrations in fish consumed in Syria. Journal of Environment Radioactivity, 2000; **49**(3): 354-352.

[53] Hoffman F.L., Hodge V.F., Folsom R.R. Po-210 radioactivity in organs of selected tunas & other marine fish. Journal of Radiation Research, 1974; **15**: 103-106.

[54] Heyraud M., Domanski P., Cherry R.D., Fasham M.J.R. Natural tracers in dietary studies: data for Po-210 & Pb-210 in decapods shrimp & other pelagic organisms in the Northeast Atlantic Ocean. Marine Biology, 1988; **97**(4): 507-519.

[55] Khan M., Wesley S. Tissue distribution of Po210 & Pb210 in selected marine species of the cost of Kudankulam, southern coast of Gulf of Mannar, India. Environment Monitoring & Assessment, 2011; **175**(1): 623-632.

[56] Alonso-Hernez C., Diaz-Asencio M., Munos-Caravaca A., Suarez-Morell E., Avila-Moreno. 137Cs & Po-210 dose assessment from marine food in Cienfuegos Bay (Cuba). Journal of Environment Radioactivity, 2002; **61**(2): 203-211.

[57] Mohamed C.A.R., Tee L.T., Zaluyun M., Zaharuddin A., Masni M.A. Activity concentrations of Po210 & Pb210 in edible tissue of fish caught at Kuala Selangor, Malaysia. Malaysian Applied Biology, 2006; **35**:67-73.

[58] Flynn W.W. The determination of low levels of Polonium-210 in environment materials. Analytical Chimica Acta, 1968; **43**:221-227.

[59] Akozcan, S., Gorgun, A.U. Variations of Po-210 and Pb-210 concentration in mussels (*Mytilus galloprovincialis*) from Didim and Izmir Bay (Turkish coast of Aegean Sea). Marine Pollution Bulletin, 2013; **68** (1-2): 152-156.

[60] Wildgust, M.A., Mcdonald, P., and White, K.N. Assimilation of Po-210 by the mussel *Mytilus edulis* from the alga *Isochrysis galbana*. Marine Biology, 2000; **136**: 49-53.

[61] Connan, O., Germain, P., Solier, L. and Gouret, G. 2007. Variations of Po-210 and Pb-210 in various marine organisms from Western English Channel: contribution of Po-210 to the radiation dose. Journal of Environmental Radioactivity, 2007; **97**(2-3): 168-188.

[62] Alam, L. and Mohamed, C.A.R. A mini review on bioaccumulation of Po-210 by marine organisms. International Food Research Journal, 2011; **18**(1).

[63] Strok, Marko and Smodis, B. Levels of Po-210 and Pb-210 in fish and molluscs in Slovenia and the related dose assessment to the population. Chemosphere, 2011; **82**(7): 970-976.

[64] Bajt, O., Ramsak, A., Milun, V., Andral, B., Romanelli, G., Scarpato, A., Mitric, M., Kupusovic, T., Kljajic, Z., Angelidis, M. and Cullaj, A. Assessing chemical contamination in the coastal waters of the Adriactic Sea using active mussel biomonitoring with *Mytilus galloprovincialis*. Marine pollution bulletin, 2019; **141**: 283-298.

[65] Jaitly, A.K. and Mishra, S.K. Campanian-Maastricthian (Late Cretaceous) veneroids (Bivalvia: Heterodonta) from the Ariyalur Group, South India. Paleoworld, 2009; **18**(4): 251-262.

[66] Checa, A.G. and Jimenez-Jimenez, A.P. Evolutionary morphology of oblique ribs of bivalves. Palaeontology, 2003; **46**(4): 709-724.

[67] Lubna, A., Nik Azlin, N.A., Afiza Suriani, S. and Mohamed, C.A.R. A study on the activity concentration of Po-210 in the marine environment of the Kapar coastal area. Journal of Tropical Marine Ecosystem, 2011; **1**(1).

[68] Carvalho, F.P., Oliveira, J.M., Alberto, G. and Vives i Batlle, J. Allometric relationships of Po-210 and Pb-210 in mussels and their application to environmental monitoring. Marine Pollution Bulletin, 2010; **60**: 1734-1742.

[69] Schell W.R. Concentration, physic-chemical states and mean residence times of Po210 and Pb210 in marine and estuarine waters. Geochimica et Cosmochimica Acta, 1977; **41**: 1019-1031.

[70] Abdullah, A.R., Tahir, N.M., Loong, T.S., Hoque, T.M. and Sulaiman, A.H. The GEF/UNDP/IMO Malacca Straits Demonstration Project: source of pollution. Marine Pollution Bulletin, 1999; **39**: 229-233.

[71] Alam L., Mohamed C.A.R. Natural radionuclide of Po-210 in the edible seafood affected by coal-fired power plant industry in Kapar coastal area of Malaysia. Environmental Health, 2011; **10**:43-52

[72] Theng T.L., Ahmad Z., Mohamed C.A.R. Activity concentrations of Po-210 in the soft parts of cockle (*Anadara granosa*) at Kuala Selangor, Malaysia. Journal of Radioanalytical and Nuclear chemistry. 2004, **262:** 485-488.

[73] Young A.K., McCubbin D., Camplin W.C. Natural Radionuclides in Seafood. Environment Report RL17/02, 2002.

[74] Rollo S.F.N., Camplin W.C., Allington D.J., Young A.K. Natural radionuclides in the UK marine environment, Radiat. Prot. Dosim., 1992; **45**, 203-209.

[75] Carvalho F. Polonium (Po-210) and lead (Pb-210) in marine organisms and their transfer in marine food chains. Journal of Environmental Radioactivity, 2011; **102**:46-472

[76] Theng T.L. and Mohamed C.A.R. Activities of Po-210 and Pb-210 in the water column at Kuala Selangor, Malaysia. Journal of Environmental Radioactivity. 2005;**80**:273-286.

[77] Khan M.F., Godwin Wesley S. Assessment of health safety from ingestion of natural radionuclides in seafoods from a tropical coast, India. Marine Pollution Bulletin, 2011; **62**(2): 399-404.

[78] Shaheed K., Somasundara S.S., Hameed P.S., Iyengar, M.A. A study of polonium-210 distribution aspects in the riverine ecosystem of Kaveri, Tiruchirappalli, India, 1997; **95**(3):371-377.

[79] Iyengar M.A.R, Kannan V., Ganapathy S., Kamath P.R. Po-210 in the coastal waters of Kalpakkam. In proceedings (second special symposium on natural radiation environment, BARC, Bombay), 1981: 227-233.

[80] Musthafa, M.S., Arunachalam, K.D., and Raiyaan, G.D. Baseline measurements of Po-210 and Pb-210 in the seafood of Kasimedu fishing harbor, Chennai, South East Coast of India and related dose to population. Environmental Chemistry and Ecotoxicology, 2019; **1**: 43-48.

Section 3

Coastal Sediments

Chapter 4

Coastal Sediment as an Ameliorant in Post-Mining Land Management

Sulakhudin and Denah Suswati

Abstract

Coastal sediment is a sediment resulting from sedimentation of eroded materials from up land through river flows that are deposited around the coast. It usually contains a lot of alkaline cations, especially Na so that it is good enough to decrease soil acidity. The use of coastal sediment must be considered carefully because it has a high level of salinity, which can inhibit plant growth and even cause death. Coastal sediment as an ameliorant can replace the role of lime in increasing the pH and base saturation of soil. Applying coastal sediment to sandy or post-gold mining soils can reduce soil acidity, increase soil CEC and soil base saturation, as well as the availability of nutrients, especially nutrients, phosphorus, potassium, calcium and magnesium. Improvement of some of these soil properties will encourage increased growth and crop yields in post gold mining land.

Keywords: ameliorant, coastal sediment, crop yield, post gold mining

1. Introduction

Apart from being one of the pillars of a country's economy, gold mining activities also contribute greatly to the rate of land degradation [1]. Furthermore, Ref. [2] explained that the result of the mining process has caused soil damage, water pollution, and the destruction of natural vegetation. If the land is left without reclamation activities, it will become critical land. Physically, the topsoil is dominated by sand particles so that the soil becomes very porous [3]. Soil that is dominated by sand causes some of the chemical properties of soil in post-mining land to be low in its ability to hold water and nutrients, have high acidity and low cation exchange capacity and base saturation [4]. Further, Ref. [5] explained that during the mining process it will destroy vegetation and some macro and micro-fauna that play a role in the processes of decomposition of organic matter and the cycle of nutrients in the soil.

Acidic soil is less able to support plant growth because some macro nutrient availability decreases if the soil pH becomes acidic. Macronutrients consisting of nitrogen, phosphate, potassium, calcium, magnesium, and sulfur, are not available or dissolve at acidic pH [6]. This is in contrast to micro-nutrients other than molybdenum which are more soluble or readily available at low pH. The level of solubility of these nutrients is often ignored in agriculture, especially by farmers who are still cultivating in traditional ways. By considering the level of solubility, a plant that is cultivated on acid soil, the nutrients that are available in large quantities are micro nutrients [7]. In fact, micro-nutrient elements that are actually only a few needed by plants are actually available in large quantities, so that they have the potential to cause poisoning to plants, for example plants become poisoned with iron (Fe) or aluminum (Al).

The high availability of micro-nutrients in acid soils also results in high bonds between soil ions. Iron, manganese and aluminum elements will bind strongly to macro nutrients, especially phosphorus. This results in the low availability of macro nutrients in acid soils. One of the ways to increase low soil pH is by liming the right amount, so that the macro nutrients needed by plants are available in large quantities and can be directly absorbed by plant roots. One of the constraints of liming is that the lime material must be imported from outside the area, so when it is needed lime is not available and the price is relatively expensive. Besides that, agricultural lime is inefficient because of its low residual level. One of the alternatives to limestone is coastal sediment which is abundant and widespread on the coast. Ref. [8] shows that coastal sediment as an ameliorant can replace the role of lime in increasing soil pH.

Utilization of coastal sediment must be carefully managed because it needs to be remembered that coastal sediment has a high level of salinity which can disrupt plant physiology and even cause death in these plants. However, it should be noted that in using coastal sediment, it is not necessary to use sediment that has been contaminated by heavy metals such as lead (Pb), mercury (Hg) and other heavy metals. Metals do not directly harm plants, but it is feared that the results of plant production if consumed will have an impact on human health [9]. Coastal sediment as an ameliorant can replace the role of lime in increasing pH and base saturation in peat soils [10]. In sandy soil/post gold mining soil which is dominated by sand fraction, application of coastal sediment can improve some of the soil properties. The addition of coastal sediment on sandy soil/land after gold mining, in addition to reducing soil acidity, can also reduce CEC, increase base saturation and the availability of cations (Ca^{2+}, Mg^{2+}, Na^{+}, K^{+}, Mn^{2+} and Fe^{2+}). Based on the description above, this chapter aims to explain the use of coastal sediment as an ameliorant in land management after gold mining for plant cultivation. This study aims to obtain the best dosage for coastal sediment to improve soil properties, growth and crop yields in post-gold mining land.

2. Charateristic of coastal sediment

Coastal Sediment is a material that is deposited by water (rivers and seas) in the form of a mixture of alluvial soil and organic matter. It is formed through the process of alluviation and collusion on land with long acid reactions, dissolving and carrying weak alkaline elements (Al, Fe, Mn) through the process of erosion and/or leaching. When alluvial/coluvial material finally settles in the sea, then marine silt deposits contain weak bases, Al, Fe, and Mn mixed with the strong bases Na, Ca, and Mg, which are contained in the sea. Weak basic elements (and their combination with weak acidic compounds) produce compounds that are "buffered", have a pH dependent ionic charge (pH dependence charge), positive (+) at low pH (acid reaction) and negative (−) at low pH. high pH (base reaction).

Buffer compounds increase the carrying capacity of nutrients, thereby increasing plant growth and productivity. Thus, the utilization of coastal sediment from fertilization/silication deposits has the potential to ameliorate acid soil. Coastal sediment acts as an ameliorant for the improvement of the physical and chemical (physico-chemical) components of the soil. The potential of coastal sediment in amelioration of physico-chemical properties needs to be assisted by ameliorant of soil biological characteristics. In agricultural practice, biological ameliorant is manure, which is rich in soil fertilizing microorganisms.

Coastal sediment is a sediment resulting from sedimentation of eroded materials from upland through river flows that are deposited around the coast. The nutrient content in coastal sediment varies greatly depending on the type of soil and the conditions of the area of origin of the sediment. Some of the chemical properties of coastal sediment taken from 3 locations (Kijing Beach; location I, Rasau Jaya Beach; location II and Muara Sungai Singkawang Beach; location III) can be seen in **Table 1**.

The results of particle analysis showed that the three coastal sediment from each location had different content of sand, silt and clay. The highest clay content was found in location II which was 56.47%. Thus the coastal sediment from Rasau Beach is suitable for application on gold ex-mining lands which in addition to increasing the pH will also improve several other soil properties. Especially to reduce the very high porosity of the used gold ex-mining soil and at the same time increase the holding capacity of soil water.

The highest pH value of coastal sediment is found in coastal sediment from Kijing Beach, which reaches 8.13 (**Table 1**), while coastal sediment from Rasau Beach and Singkawang River Estuary is only 7.72 and 7.14, respectively. Based on the pH data, coastal sediment from Kijing Beach can be used on all types of soil in the West Kalimantan Province with a relatively small amount compared to coastal sediment from other locations to raise the pH. Based on the nutrient content, each coastal sediment from the three locations has different advantages. Coastal sediment from location I had the highest total nitrogen content of 7.26%, while at locations II and III were 0.98 and 0.27%, respectively. Coastal sediment from location II has the highest P content than coastal sediment at locations I and III. The P content at location II was 10.24 ppm, while at locations I and III were 3.45 and 9.65 ppm, respectively. Coastal sediment from location II has the highest potassium content of 5.01 cmol (+) kg^{-1}, while at locations I and III are 1.71 and 3.76 cmol (+) kg^{-1}, respectively.

Soil chemical parameters	Coastal sediment		
	Location I	Location II	Location II
Tekstur			
Sand (%)	10,20	5,31	1,31
Silt (%)	51,85	38,22	44,79
Clay (%)	37,95	56,47	53,90
pH	8,13	7,72	7,14
C-organic (%)	1,96	1,18	2,05
N-total (%)	7,26	0,98	0,27
P Bray I (ppm)	3,45	10,24	9,65
K (cmol(+)kg^{-1})	1,71	5,01	3,76
Ca (cmol(+)kg^{-1})	14,62	65,10	11,44
Mg (cmol(+)kg^{-1})	1,73	10,24	3,81
Na (cmol(+)kg^{-1}	2,65	36,03	34,85
CEC ($cmol^{(+)}kg^{-1}$)	15,33	15,82	11,55
Base saturation (%)	>100	>100	>100

Table 1.
Characteristics of coastal sediment from several locations in West Kalimantan.

Coastal sediment from locations II and III has a higher sodium content than coastal sediment from location I. At location I the Na content is only 2.65 cmol (+) kg^{-1}, while at locations II and III the coastal sediment contains Na respectively 36.03 and 34.85 cmol (+) kg^{-1}, respectively. The Na content of 15 times from coastal sediment in location I is dangerous because Na has a bad effect on several soil properties [11]. Thus, in the use of coastal sediment from locations II and III, it is necessary to reduce Na by washing so that the Na content is lower.

The high Ca content of coastal sediment at location II (65.10 cmol (+) kg^{-1}) is not only a source of nutrients but also to maintain the balance of nutrients in the soil [12]. The base saturation (BS) data of coastal sediment is more than 100% so that the application of coastal sedimentis expected to increase soil pH and BS. Based on the comparison data of several chemical properties of the soil, coastal sediment from location I, namely Kijing beach, is the best coastal sediment as an alternative to lime compared to coastal sediment from locations II and III.

3. Characteristic of post-mining land

Land at the post-mining site without a permit has suffered considerable damage. Soil damage from physical, chemical and biological characteristics causes the soil to be unable to support optimal plant growth, so that this land is left to become abandoned land [13]. The current condition of the post-mining land without permits is over grown with shrubs with that grass as the dominant plant with several basins from the former mining activity.

Some of the chemical and physical properties of the soil used in the study are listed in **Table 2**. These characteristics are properties of the soil in post-gold mining land in Mandor Sub-District, Landak Regency, West Kalimantan Province. These soil properties illustrate some of the problems in the land after the gold mining from the physical and chemical properties of the soil. The soil texture class is classified as sand because soil particles are dominated by the sand fraction which reaches 91.53%, while the silt and clay fractions are only 8.11% and 0.36%, respectively [14]. The percentage of the sand fraction that reaches more than 90% characterizes sandy soils or in mining terms it is called tailings.

Soil whose particle fraction is dominated by sand has a high permeability, this will cause the leaching rate of nutrients in the soil to be very high [15]. As a result, the availability of the nutrient is low to very low. **Table 2** shows some properties of soil in post gold mining at Mandor Sub District i.e. the total nutrient content of N (0.02%), Ca (0.13 cmol (+) kg^{-1}), Mg (0.38 cmol (+) kg^{-1}) and Na (0.09 cmol (+) kg^{-1}) available is very low, while P and K of 6.64 ppm and 0.15 cmol (+) kg-^{1} respectively are classified as low.

Potassium available in the soil in people's post gold mining land of 0.15 cmol (+) kg^{-1} is low. Generally the sandy soil is sufficiently K, but most of it is only in the form not yet available to plants, K is still in primary minerals such as feldspar and mica in sand particles. The very low nutrient content in the soil in post community gold mining land as mentioned above is also caused by the low nutrient binding sand soil, which is reflected in the very low of CEC value of 3.54 cmol (+) kg^{-1}.

The very low value of the CEC on this soil is due to several things, including: (1) The low clay fraction (0.36%) which is a source of negative charges; (2) The organic matter content is very low, which is reflected in the low C-organic value, namely 0.01%. Very low soil organic matter can be caused by the fast rate of decomposition of organic matter in sandy soils due to the high temperature and aerobic atmosphere [16]. The soil pH value in the post-gold mining land area of 5.63 is classified as

	Location I		Location II	
Soil properties	**Value**	**Level**	**Value**	**Level**
pH H_2O 1:2	4,9	Acid	5,63	slightly acid
pH KCl 1:2	4,3	Very Acid	4,21	Very Acid
C-Org (%)	2,59	Moderate	0,01	Very low
N Total (%)	0,31	Moderate	0,02	Very low
P Bray I (ppm)	10,19	High	6,64	Low
Ekstract NH_4OAc 1 N pH 7				
K (cmol(+)kg^{-1})	0,23	Low	0,15	Low
Ca (cmol(+)kg^{-1})	1,68	Very low	0,13	Very low
Mg (cmol(+)kg^{-1})	1,05	Moderate	0,38	Very low
Na (cmol(+)kg^{-1}	0,26	Low	0,09	Very low
CEC (cmol(+)kg^{-1})	10,64	Low	3,54	Very low
Base saturation (%)	29,61	Low	21,74	Low
Texture				
Sand (%)	86	Sand	91,53	Sand
Silt (%)	12		8,11	
Clay (%)	2		0,36	

Note: Marking according to the Soil Research Institute (2005): Location I is in Singkawang Sub-District and location II is in Mandor Sub-District.

Table 2.
Soil characteristics of post gold mining land in some location.

slightly acidic. Soil pH value will be a limiting factor for plant cultivation, so that the growth of plants is less than optimal [17]. One of the alternatives to increase the pH by applying coastal sediment. Besides being able to increase the pH and availability of several nutrients, it can improve some of the physical properties of the soil in post gold mining land. This is because the coastal sediment contains 37.95% clay.

The results of the analysis of several soil properties indicate that the soil in the post-mining area of gold without a permit in Central Singkawang District has decreased its fertility. This is indicated by the very low ability of the soil to bind nutrients and water. The ability of soil to bind water and soil nutrients can be seen based on the very low value of the cation exchange capacity (CEC), namely 4.74 cmol (+) kg^{-1} (**Table 2**). In addition, the low fertility level can also be seen from the texture of the soil, namely sand. Soil whose mineral fraction is dominated by sand will cause the ability to store water and nutrients to be low because sand has a low negative charge [18]. Soil whose particle fraction is dominated by sand has a high permeability, this will cause the leaching rate of nutrients in the soil to be very high. As a result, the availability of nutrients becomes low to very low. The very low CEC value in this soil is caused by several reasons, including: (1) it does not contain clay fraction (0.00%) which is a source of negative soil charge; (2) The organic matter content is very low as indicated by the low C-organic value, namely 0.21%. Very low soil organic matter can be caused by the fast rate of decomposition of organic matter in sandy soils due to the high temperature and aerobic atmosphere. The results of the analysis in **Table 1** show that the organic matter in the sand has been further decomposed with a C/N ratio value of 8.35.

The soil pH value in the post-gold mining land area of 4.9 is considered acidic. Soil pH value will be a limiting factor for plant cultivation because in acid soils some nutrients are not available, for example K, Ca and Mg so that they cannot provide optimal nutrients for plant growth [19].

Community gold mining produces mercury as the main pollutant that will threaten the sustainability of the ecosystem. Mercury can damage the environment because of its low solubility in water and is easily absorbed and accumulated in the tissues of organisms through bioaccumulation and biomagnification processes [20]. Mercury levels in 4–5 year old gold mining land is 0.020 ppm, 6–10 year old gold mining land is 0.050 ppm and 0.042 year old gold mining land is 0.042 with an average grade of 0.037 ppm. Mercury and its derivatives are one of the deadliest pollutants in the history of human civilization [21].

4. The role of coastal sediment in increasing growth and crop yields in post-mining land

The concentration of several nutrients in the sorghum plant tissue due to coastal sediment addition can be seen in **Table 3**. The variations of N, P, K, Ca and Mg contents in the sorghum plant from different provision of coastal sediment were considerable at several doses level, especially if compared with control. The Ca concentration in sorghum plant that were given coastal sediment at all doses showed increased compared to control (**Figure 1**). The increasing of concentration of Ca in the sorghum crop due to coastal sediment addition caused by coastal sediment many contain Ca. Research result of [22] indicates that the coastal sediment contains Ca of 14.62 cmol (+) kg^{-1}.

The provision of coastal sediment is able to increase the concentration of P in the sorghum plant, the highest concentration of P at treatment of coastal sediment addition at a dose 60 t ha^{-1}. The concentration of P elements on all the addition of coastal sediment is significant difference with control. This is due to the addition of coastal sediment can increase soil pH, according to [23] provision of coastal sediment increase significantly soil pH because it contained high cations. The higher soil pH value then the availability of P will be higher so that sorghum plant can absorb more P elements.

Table 3 shows the uptake some nutrients of sorghum plants in post gold mining land. Absorption of nutrients at all doses of coastal sediment application

Treatment	Uptake nutrients				
	N	P	K	Ca	Mg
			(mg)		
Rates of coastal sediment					
0 t ha^{-1}	146.9 b	0.8 c	153.5 b	25.6 c	19.6 b
20 t ha^{-1}	390.2 a	1.4 ab	421.4 a	87.3 a	44.3 a
40 t ha^{-1}	296.4 ab	1.2 b	262.2 b	52.2 bc	29.8 ab
60 t ha^{-1}	400.7 a	1.5 a	413.0 a	83.4 ab	45.5 a
80 t ha^{-1}	239.6 ab	1.3 ab	292.5 ab	61.6 ab	31.1 ab

Description: Numbers followed by the same letters in the same column indicate no significant differences at the Duncan test at 5% level of significance.

Table 3.
Effect of coastal sediment application on some uptake nutrient by sorghum.

was higher than control. Uptake of N nutrients due to giving of coastal sediment ranges from 239.6–400.7 mg, while the control of only 146.9 mg of N was absorbed by the Sorghum plant. The phosphorus and calcium uptake in tissue of sorghum as measured after harvesting were significantly increased by coastal sediment application.

The provision of coastal sediment of dosage at 60 t ha^{-1} increased the highest P uptake by sorghum. Application of coastal sediment at dose 60 t ha^{-1} increase the highest uptake of P with a value of 1.5 mg, when compared with the absorption of P without the provision of coastal sediment then the absorption of P increased by 46.67%. Ref. [24] states that on acid soil increased pH will increase the absorption of P plants. **Table 2** also showed that the addition of coastal sediment at a dose of 60 t ha^{-1} can increase the highest Mg uptake, which is 45.5 mg. Increased absorption of Mg in the application of coastal sediment dose at 60 t ha^{-1} compared to a control of 56.92% (**Figure 2**).

The effect of coastal sediment application on the yield of sorghum crops is known from the number of seeds per plant (NSP), weight per plant (WPP), and weight per 100 seed (W100S). **Table 4** shows that the provision of coastal sediment at all dosages differs significantly against the number of seeds per plant than the control. The NSP value of coastal sediment addition ranged from 1362 to 2082 seeds,

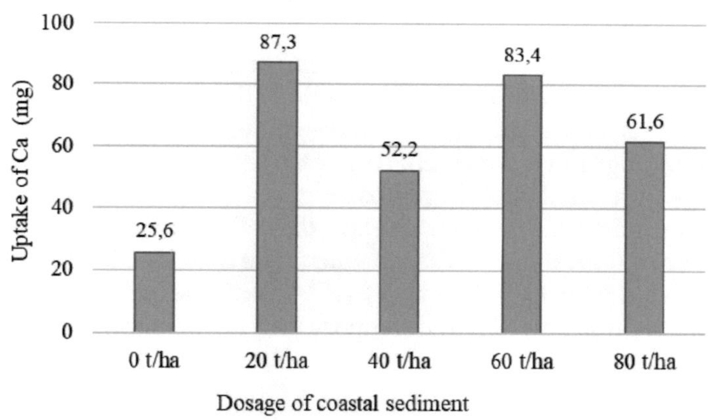

Figure 1.
Plant Ca uptake at several dosage of coastal sediment addition.

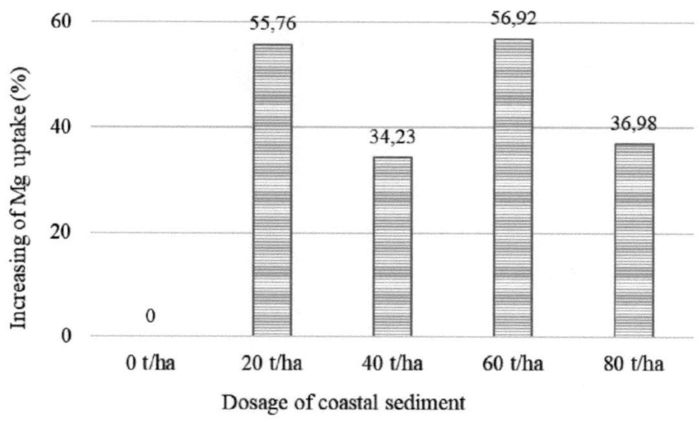

Figure 2.
The percentage of increased Mg uptake due to the provision of coastal sediment at several doses.

Treatment	NSP	WPP	W100S
Rates of coastal sediment			
0 t ha^{-1}	683 c	14.4 c	2.11 b
20 t ha^{-1}	1711 ab	37.4 ab	2.4 ab
40 t ha^{-1}	1362 b	32.2 b	2.47 a
60 t ha^{-1}	2082 a	48.9 a	2.59 a
80 t ha^{-1}	1881 ab	43.3 ab	2.50 a

Description: Numbers followed by the same letters in the same column indicate no significant differences at the Duncan test at 5% level of significance.

Table 4.
Effect of ameliorant on some yield sorghum parameter.

while the control only has an NSP of 683 seeds. Likewise, for WPP parameters, the treatment of coastal sediment addition at all doses is significantly different with the control. It increases in the amount of weight per plant between 55.28–70.55%.

The weight per 100 seed parameters also shows an increase in sorghum plant with addition of coastal sediment. The weight increase per 100 seeds appears to be a distinct significant start of coastal sediment application at doses of 40 t ha^{-1}, while at doses of 20 t ha^{-1} was not differ from the control. **Table 4** shows that on all three parameters, the provision of coastal sediment doses 60 t ha^{-1} has the highest value. Then at a higher dose, i.e. 80 t ha^{-1} precisely the three parameters indicate the decline. This means the dosing of coastal sediment for the sorghum plant in the post gold mining land at a dose above 60 t ha^{-1} began to decrease the yield of sorghum crops. Suspected with the provision of coastal sediment that is too high will interfere with the balance of nutrients in the soil, especially because of the influence of the sodium elements are too much. The coastal sediment contains Na which is quite high, namely 3.24 cmol (+) kg^{-1}. One of the bad influences of Na is that it can reduce the absorption of other positively charged nutrients, such as K, Ca and Mg [25].

5. Final remarks

The post gold mining land has the potential for the development of crops production with the provision of coastal sediment ameliorant. It can increase the uptake of nutrients N, P, K, Ca and Mg, as well as crop results. The optimum dose of coastal sediment giving to the sorghum plant in the post gold mining land is 60 t ha^{-1}.

Author details

Sulakhudin* and Denah Suswati
Tanjungpura University, Pontianak, Indonesia

*Address all correspondence to: sulakhudin@faperta.untan.ac.id

References

[1] Eludoyin AO, Ojo AT, Ojo TO, Awotoye OO. Effects of artisanal gold mining activities on soil properties in a part of southwestern Nigeria. Cogent Environ. Sci. Jan. 2017;**3**(1):1305650. DOI: 10/gg9hgr.

[2] K. Peprah, "Land degradation is indicative: proxies of forest land degradation in Ghana," *J. Degraded Min. Lands Manag.*, vol. 3, no. 1, Art. no. 1, Oct. 2015, doi: 10/gg9hmt.

[3] M. Nurcholis, A. Wijayani, and A. Widodo, "Clay and organic matter applications on the coarse quartzy tailing material and the sorghum growth on the post tin mining at Bangka Island," *J. Degraded Min. Lands Manag.*, vol. 1, no. 1, Art. no. 1; http://web.archive.org/web/20200902035806/https://jdmlm.ub.ac.id/index.php/jdmlm/article/view/9; http://web.archive.org/web/20200902035816/https://jdmlm.ub.ac.id/index.php/jdmlm/article/view/9/14, Oct. 2013.

[4] S. Mastur, D. Suswati, and M. Hatta, "The effect of ameliorants on improvement of soil fertility in post gold mining land at West Kalimantan," *J. Degraded Min. Lands Manag.*, vol. 4, no. 4, p. 873, 2017, doi: 10.15243/jdmlm.2017.044.873.

[5] Rieder SR, Frey B. Methyl-mercury affects microbial activity and biomass, bacterial community structure but rarely the fungal community structure. Soil Biology and Biochemistry. Sep. 2013;**64**:164-173. DOI: 10/f44v5n.

[6] White PJ, Brown PH. Plant nutrition for sustainable development and global health. Annals of Botany. Jun. 2010;**105**(7):1073-1080. DOI: 10/drb4hs.

[7] S. S. Dhaliwal, R. K. Naresh, A. Mandal, R. Singh, and M. K. Dhaliwal, "Dynamics and transformations of micronutrients in agricultural soils as influenced by organic matter build-up: A review," *Environ. Sustain. Indic.*, vol. 1-2, p. 100007, Sep. 2019, doi: 10/ghchnk.

[8] "Suswati et al_2015_Use of ameliorants to increase growth and yield of maize (*Zea mays* L.pdf." Accessed: Aug. 29, 2020. [Online]. Available: https://journal.unila.ac.id/index.php/tropicalsoil/article/download/198/197.

[9] Jaishankar M, Tseten T, Anbalagan N, Mathew BB, Beeregowda KN. Toxicity, mechanism and health effects of some heavy metals. Interdisciplinary Toxicology. Jun. 2014;7(2):60-72. DOI: 10/gcsgpf.

[10] Suswati D, Sunarminto BH, Indradewa D. Use of ameliorants to increase growth and yield of maize (*Zea mays* L.) in peat soils of West Kalimantan. J. Trop. Soils. 2015;**19**(1):35-41. DOI: 10.5400/jts.2014.v19i1.35-41

[11] Phogat V, Mallants D, Cox JW, Šimůnek J, Oliver DP, Awad J. Management of soil salinity associated with irrigation of protected crops. Agricultural Water Management. Jan. 2020;**227**:105845. DOI: 10/ghd354.

[12] P. J. WHITE and M. R. BROADLEY, "calcium in plants," Annals of Botany, vol. 92, no. 4, pp. 487-511, Oct. 2003, doi: 10/bv8cb2.

[13] Sheoran V. A. S. Sheoran, and P. Poonia, "Soil Reclamation of Abandoned Mine Land by Revegetation: A Review,". 2010;**3**:21

[14] Silva SHG et al. Soil texture prediction in tropical soils: A portable X-ray fluorescence spectrometry approach. Geoderma. Mar. 2020;**362**:114136. DOI: 10/ghd3z4.

[15] S. Tahir and P. Marschner, "Clay Addition to Sandy Soil Reduces Nutrient Leaching—Effect of Clay Concentration and Ped Size," *Commun. Soil Sci. Plant Anal.*, vol. 48, no. 15, pp. 1813-1821, Aug. 2017, doi: 10/ghd4cb.

[16] A. Campos C., G. Suárez M., and J. Laborde, "Analyzing vegetation cover-induced organic matter mineralization dynamics in sandy soils from tropical dry coastal ecosystems," *CATENA*, vol. 185, p. 104264, Feb. 2020, doi: 10/ghd32z.

[17] Neina D. The role of soil pH in plant nutrition and soil remediation. Applied and Environmental Soil Science. Nov. 03, 2019. DOI: https://doi.org/10.1155/2019/5794869 (accessed Sep. 04, 2020)

[18] Xie L, Li J, Liu Y. Review on charging model of sand particles due to collisions. Theoretical and Applied Mechanics Letters. Apr. 2020;**10**(4):276-285. DOI: 10.1016/j.taml.2020.01.047

[19] Gentili R, Ambrosini R, Montagnani C, Caronni S, Citterio S. Effect of soil pH on the growth, reproductive investment and pollen Allergenicity of Ambrosia artemisiifolia L. Frontiers in Plant Science. Sep. 2018;**9**. DOI: 10/ghd3xx.

[20] Sierra MJ, Rodríguez-Alonso J, Millán R. Impact of the lavender rhizosphere on the mercury uptake in field conditions. Chemosphere. Nov. 2012;**89**(11):1457-1466. DOI: 10.1016/j.chemosphere.2012.06.017

[21] M. Gochfeld, "Cases of mercury exposure, bioavailability, and absorption," Ecotoxicol. Environ. Saf., vol. 56, no. 1, pp. 174-179, Sep. 2003, doi: 10/dsvj5h.

[22] "Suswati et al_2015_Effect of Ameliorants on Nutrient Uptake and Maize Productivity in Peatlands.pdf." Accessed: Aug. 29, 2020. [Online]. Available: http://www.msss.com.my/mjss/Full%20Text/vol19/10-Suswati.pdf.

[23] Arief FB, Gafur S, Sagiman S, Aspan A. Characteristics of coastal sediment from three different sites and their potential as the ameliorant of peat soil in West Kalimantan. in *IOP Conference Series: Earth and Environmental Science*. 2019;**393**(1):012033. DOI: 10.1088/1755-1315/393/1/012033.

[24] Fageria NK, Nascente AS. Chapter six - Management of Soil Acidity of south American soils for sustainable crop production. In: *Advances in Agronomy*. Vol. 128. D. L. Sparks: Ed. Academic Press; 2014. pp. 221-275

[25] B. Çalişkan and A. C. Çalişkan, "Potassium Nutrition in Plants and Its Interactions with Other Nutrients in Hydroponic Culture," *Potassium - Improv. Qual. Fruits Veg. Hydroponic Nutr. Manag.*, Dec. 2017, doi: 10/ghd37x.

Section 4

Coastal Ecosystems

Chapter 5

Enterococcus Present in Marine Ecosystems and Their Potential to Degrade Azo Dyes

Ganiveth María Manjarrez Paba and Rosa Baldiris Ávila

Abstract

Azo dyes are frequently used at an industrial level to restore the color of raw materials once it has faded away, make an original color more vibrant or with the purpose of giving a material a different color that is considered more attractive. These processes however, have a negative impact on the environment, evidenced in colored wastewater that is subsequently dumped into water bodies, causing disruptions in the natural balance of ecosystems and deteriorating human health. Traditional strategies for the treatment of effluents contaminated with azo dyes are limited to physical and chemical processes that have a high energy and economic cost. For these reasons, current challenges are focused on the use of microorganisms capable of transforming dyes into less toxic products. This chapter will present a description of the main characteristics of azo dyes and the different methods used for their treatment, with special emphasis on the benefits associated with biological treatment. Likewise, it will provide relevant information about *Enterococcus* and show its potential to degrade azo dyes.

Keywords: Enterococcus, marine, azo dyes

1. Introduction

Annually, more than a million tons of synthetic dyes are produced around the world for use in the leather, textile, pharmaceutical, food, cosmetic, paint, plastic and paper industries [1], of which, at least 60% represent azo dyes [2]. In addition to being recalcitrant towards various degradation processes [3], azo dyes produce dangerous chemical substances such as aromatic amines, known for their toxic, allergenic, carcinogenic and mutagenic effect on living organisms [4].

The impact of azo dyes on the environment is related to the enormous amounts of hazardous waste associated with industrial processes, which in most cases, is then released directly to water bodies without proper treatment. A further aggravating factor is that due to the inability of at least 35% of azo dyes to adhere to substrates, heavy metals have been incorporated during the dyeing process, and these act as mordants, favoring the fixation of the dye [5].

Colorants associated with metals such as copper, cobalt and especially chromium, are difficult to degrade and represent an important source of environmental contamination due to their increased presence in organic load. They generate adverse and irreversible eco-toxicological effects, bioaccumulation phenomena and biomagnification in flora and aquatic fauna and alteration of biogeochemical cycles [6].

This powerful metal-dye complex has carcinogenic and mutagenic properties for humans exposed to effluents contaminated with dyes. It can lead to skin cancer due to photosensitization, photodynamic damage, allergic contact dermatitis, renal, reproductive, hepatic, cerebral dysfunction, irritation of the respiratory tract and asthma [7].

Traditionally, physicochemical methods have been used to treat effluents contaminated with azo dyes, but their high economic and energy cost and the environmental effects associated with their use have changed the focus, in recent years, on the use of microorganisms. These are successful biological alternatives due to their survival properties, adaptability, enzymatic activity and chemical structure. Additionally, hybrid technologies have been developed, which are able to take the best of each technology and surpass the limitations of current conventional treatments [8].

Enterococcus sp. are gram-positive cocci, facultative anaerobes capable of growth in environments with low nutrient concentrations, persistent temperature fluctuations, and are resistant to desiccation, UV radiation, freezing, pH changes, high salinity and predation [9]. According to phylogenetic studies, this genus includes 50 species of clinical and environmental importance [10].

The environmental importance of *Enterococcus* sp. has to do with the fact that, since 1986, the US Environmental Protection Agency included them as part of the parameters for evaluating the quality of marine waters. Likewise, the World Health Organization considered them more important than thermotolerant coliforms, due to their ability to resist the physical and chemical conditions of seawater, and for being an excellent indicator for waters impacted by fecal contamination [11].

While international organizations consider *Enterococcus* sp. as indicators of fecal contamination of marine waters, another relevant aspect and a novelty of this article has to do with taking into consideration beaches as well. Beaches are complex ecosystems where there is a dynamic of continuous transport between water and sand. For this reason, it was considered decisive not only to evaluate *Enteococcus* sp. in water, but also in beach sand, where animal feces and residues generated by anthropogenic activities are generally found [12].

As a contribution to this discussion and element for further research, this article presents a review of the potential of *Enterococcus* to become an optimal biological alternative in the treatment of effluents contaminated with Azo dyes. This is due to its ability to survive in aquatic environments with adverse environmental conditions, its development of multi-resistance mechanisms for antibiotics and heavy metals, as well as its enzyme systems associated with the degradation of dyes.

1.1 The potential of bacteria for the degradation of azo dyes

For the degradation of azo dyes, bacteria have an efficient enzymatic system that allows them to carry out a series of catabolic activities, with azoreductase and laccase enzymes being responsible for the transfer of electrons to the azo bond of the dye and the production of aromatic amines [13].

The mechanism of degradation by azoreductase enzymes consists of two phases. The first, called the reducing phase, begins with the cleavage of the azo bond (-N=N-) by catalyzed reduction of the enzyme under anaerobic/anoxic or microaerophilic conditions, where NADH molecules, derived from carbohydrate metabolism are used as electron donors [13]. In the second phase, as a result of this division, relatively simple intermediate aromatic amines are generated, which are deaminated or dehydrogenated by bacteria through aerobic processes under aerobic conditions, which leads to complete degradation of azo dyes [14].

Laccases, on the other hand, are copper oxidases that degrade dyes in the presence of oxygen through mechanisms that involve direct or indirect oxidation using redox mediators to accelerate the reaction. This involves the removal of a hydrogen atom from the hydroxyl and amino groups, replacing it with phenolic substrates and aromatic amines [15].

Bacterial action in the degradation of azo dyes is increased due to their ability to act through consortiums or synergistic associations that act as biological inducers. The union of the catabolic functions of each microorganism makes them even more useful alternatives to improve the discoloration rate of effluents contaminated with dyes, as they have greater resistance to abiotic conditions and lower rates of enzyme inactivation, especially in large-scale operations [16].

1.2 Enterococcus, potentially degrading bacteria of the complex azoic dyes - heavy metals

One of the bacteria identified as an effective biological alternative for the removal of metal-dye synergy is *Enterococcus* sp., recognized for its ability to thrive in environments with low nutrient concentrations, persistent to temperature fluctuations and resistant to desiccation, UV radiation, freezing, pH changes, high salinity and predation. Furthermore, they are considered catabolically versatile microorganisms, capable of using a wide range of unusual substrates as carbon source [17].

For a long time, the environmental importance of *Enterococcus* sp. had to do with it being an excellent indicator of fecal contamination in waters [11]. However, new potential uses of this microorganism have emerged recently. It can be exploited for the benefit of the environment, such as for its ability to metabolize xenobiotics, among which are azo dyes, and it has an affinity to bind and resist heavy metals. Furthermore, the genome of these bacteria also reveals the presence of phages, which in large-scale industrial processes could be useful in improving its general bioremediation capacity and could also prove to be a viable option in transferring the ability to degrade azo dyes to other *Enterococcus* through genetic engineering from hybrid strains [18].

The ability of *Enterococcus faecalis* to metabolize azo dyes is associated with the presence of the azoA gene. This encodes the production of the aerobic azoreductase enzyme, which is not secreted outside the cell, has a wide substrate specificity, requires flavin mononucleotide (FMN) as a cofactor and uses NADH as an electron donor [19].

The ATCC 6569 *Enterococcus faecium* strain possesses the enzyme azoreductase (AzoEf1) which shares 67% identity with the azoreductase of *Enterococcus faecalis* (AzoA). However, there are differences related to coenzyme preference, residues associated with FMN binding, substrate specificity, and specific activity. The AzoEf1 sequence is found in GenBank: GQ479040.1 [20].

Enterococcus casseliflavus, by the action of an enzyme which acts similarly to that of azoreductase, is not only able to discolor a wide range of azo dyes under micro-aerophilic conditions, but also to catabolize by desulfonation and deamination the intermediaries generated as a consequence of the reductive cleavage. The genome of this microorganism also reveals the presence of regulatory systems possibly involved in the biodegradation of aromatic contaminants [21].

Enterococcus gallinarum offers an effective ecological alternative for the remediation of environments contaminated with structurally complex and recalcitrant azo dyes such as Reactive Red 35. This is done through enzymatic mechanisms that involve the presence of oxidoreductases, such as laccases, tyrosinases and azoreductases, under a wide range of pH, temperature and with high concentration of salinity. Therefore, its use on a large scale is recommended by using a suitable microaerophilic-aerobic sequential bioreactor [22].

The binding affinity of *Enterococcus* sp. to heavy metals has been attributed to the capsular polysaccharide, which contains different monomers such as glucose, galactose, mannose and fructose, and is capable of participating in the redox reaction of remediation processes of waters contaminated with heavy metals and dyes [23]. Recently, these monomers have been used for the synthesis of silver nanoparticles (AgNP) that, combined with advanced oxidation processes (AOP), have shown good results in the degradation of azo dyes such as methyl orange and Congo red [24].

In relation to metal removal, *Enterococcus faecalis* uses mechanisms such as copper transporting ATPases, present in the inner membrane, which not only work for the homeostasis of this metal but also to resist high concentrations of nickel, mercury, cadmium, lead and copper [25].

2. Methodology

Taking as reference the results of the Environmental Quality Program of Tourist Beaches [26], samples of water and sand were taken at Bocagrande beach in Cartagena Colombia, taking as reference points the areas with the highest concentration of users and suspected of contamination from a source point of marine water dumping.

To search for *Enterococcus* sp., fifty-four (54) samples were taken: 24 of water and 30 of sand, at Bocagrande beach, in Cartagena, Colombia. The standardized protocols of the Environmental Quality Program of Tourist Beaches in the Colombian North Caribbean were taken as reference [26]. The samples were transported at 4°C to the Environmental Microbiology laboratory of the University of Cartagena-Colombia, to be processed in a period of 8 to 24 hours.

The samples were processed through the membrane filtration method. In the case of the sand samples, 10 g of these were diluted in 90 mL of deionized water, the supernatant being considered as a filterable material. Filters were transferred to Slanetz & Bartley agar and incubated for 48 hours at 35 ± 0.5° C [27]. After the incubation period, the colony count was performed and the results were reported in CFU/100 ml of sample. To confirm the identification of *Enterococcus* sp., colonies were placed on blood agar supplemented with 5% defibrinated lamb blood and stored in BHI broth supplemented with 20% glycerol at −80°C.

Biochemical tests were carried out for the confirmation of the genus *Enterococcus* sp. such as: catalase, hydrolysis of esculin bile, reduction of potassium tellurite, vogues proskauer and growth tolerance in the presence of 6.5% NaCl at 9.6 pH. Gelatinase production was evaluated using Columbia agar plates containing gelatin (30 gr/l) incubated at 37°C for 48 hours. A clear area around the colonies was considered as a positive result [28]. The strains that were used as controls for all tests were: *E. fæcalis* ATCC 29212, *K. pneumoniae* 700603, *E. coli* ATCC 25922 and *S. aureus* ATCC 25923.

For the identification of microorganisms by MALDI-TOF Mass Spectrometry, it was necessary to extract ribosomal proteins, using the formic acid extraction method. The analysis of the mass spectra was performed using a Microflex LT mass spectrometer, using the MALDI Biotyper 3.4 software package from Bruker Daltonik [29]. The interpretation of the results was based on accepting scores between 2.0 and 1.7 for the identification of genus and species. Scores below 1.7 were considered unreliable.

Advanced proteomic computational techniques were used; Their advantages lie not only in the speed and economic cost of the process, but also in the effectiveness for the structural and functional analysis of the azoreductase enzymes present in *Enterococcus faecalis*; their properties can be used to potentiate its industrial applications [30].

3. Results and discussion

Of the 54 samples analyzed by the membrane filtration method, in 36 samples (64.86%) colony growth was obtained on Slanetz and Bartley agar, presumptive of the genus *Enterococcus* [27]. Greater data dispersion was observed in the results of the samples of sand, with values ranging from 0 CFU / gr to 200 CFU /gr. The results in the water samples ranged from 6 CFU / 100 mL to 47.3 CFU/100 mL, as shown in **Figure 1**. The higher percentage of positive samples in sand could be explained by the relationship between the ability of *Enterococcus* sp. to form biofilm and to persist in hostile environments such as beach sand [31].

This is due to the ability of microorganisms to adhere to particulate materials, from which they obtain protection against predation and adverse environmental conditions such as: solar radiation, pH, temperature or bioavailability of nutrients. At the same time, this provides them with a food source that allows them to survive for longer periods, favoring their multiplication [32].

Confirmatory biochemical tests were performed to the 36 samples in which growth of presumptive colonies of the genus *Enterococcus* sp. was obtained. These included hydrolysis of esculin, absence of gelatinase activity, growth in 6.5% NaCl and catalase. 100% of the strains that were identified as presumptive for *Enterococcus* sp. due to their growth on Slanetz & Bartley agar showed the ability to hydrolyze esculin in the presence of bile salts. According to literature, this positive result confirms the presence of *Enterococcus* sp. [33].

Another feature that characterizes *Enterococcus* sp. is its ability to grow at a concentration of 6.5% NaCl, and its inability to break down hydrogen peroxide into water and oxygen through the enzyme catalase. However, the results obtained in these tests do not coincide with this fact about *Enterococcus* sp. reported in literature. Rather, they suggest the presence of bacteria of the genus Streptococcus sp. due to the resistance of some strains to growth in high concentrations of NaCl, or bacteria of the genus *Staphylococcus* sp. in the case of those strains whose catalase results were positive [34].

Regarding the tests for the determination of species, the enzymatic activity of gelatinase was not expressed in any of the 36 strains identified as presumptive for *Enterococcus* sp. due to growth in Slanetz & Bartley agar; This indicates loss of the gelE phenotype, which according to literature is produced in *E. faecalis* isolates [35].

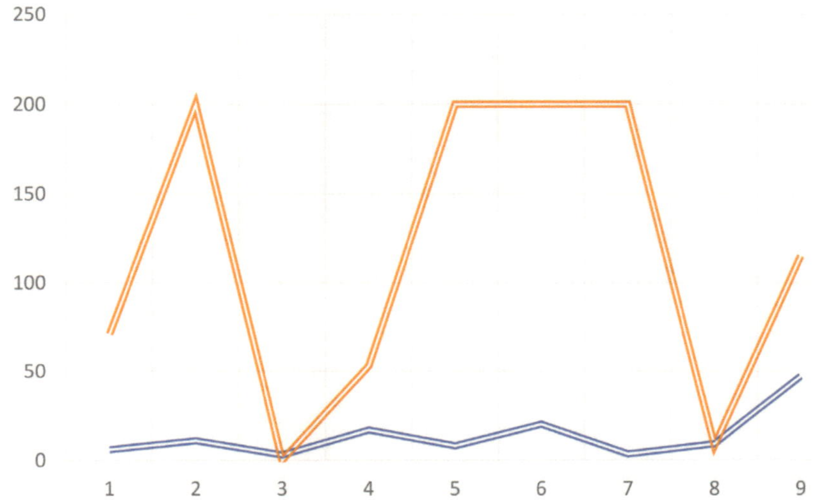

Figure 1.
Enterococci spp. levels in water and sand. Blue line: Enterococcus *in water, Orange line:* Enterococcus *in sand, Y axis: colony forming units, X axis: Weeks of monitoring.*

On the other hand, the ability to reduce potassium tellurite is one of the tests that allows differentiation of *Enterococcus faecalis* from other *Enterococcus* species. Seven (7) strains isolated from water samples and 24 strains isolated from sand samples were positive for tellurite reduction. Negative tellurite strains may suggest the presence of *Enterococcus faecium*, *Enterococcus durans*, *Enterococcus gallinarum*, or *Enterococcus casseliflavus* [36].

Taking into account the discrepancy in the results obtained, high-precision confirmatory tests were performed using matrix-assisted laser ionization mass spectrometry or MALDI TOF. Unique mass peaks are considered specific biomarkers for each genus and species. In species discrimination, MALDI-TOF MS allowed the identification of 12 strains belonging to three different species of *Enterococcus* sp. as follows: *E. faecalis* (8/32), *E. faecium* (3/32), *E. hirae* (1/32). The spectrogram of the identified *Enterococcus* is shown in **Figure 2**.

The in-silico analysis showed a low amount of cysteine residues and a high amount of aliphatic amino acids in the primary structure, which indicated that Azoreductase (2HPV) is some intracellular proteins. The hydrophobicity condition of cysteine suggests that the enzyme is nonpolar and hydrophilic in nature. The presence of a high percentage of α helices indicates that Azoreductase (2HPV) is considered thermostable, as shown in **Figure 3**.

More than 90% of the amino acids were located in the allowed region of the Ramachandran graph, which indicates their stability in nature. The results obtained by SAVES showed that the enzyme have stable crystallography and the SWISS-MODEL QMEAN, ANOLEA and ERRAT analyzes confirmed their good quality. The structural analysis established that Azoreductase (2HPV) have better thermal stability and a superior quality model than other enzymes degrading dyes such as peroxidases and laccases [37], as shown in **Table 1**.

The STRING analysis (protein –protein) identified that the proteins that interact with the azoreductase studied had an unknown 3D structure. However, the

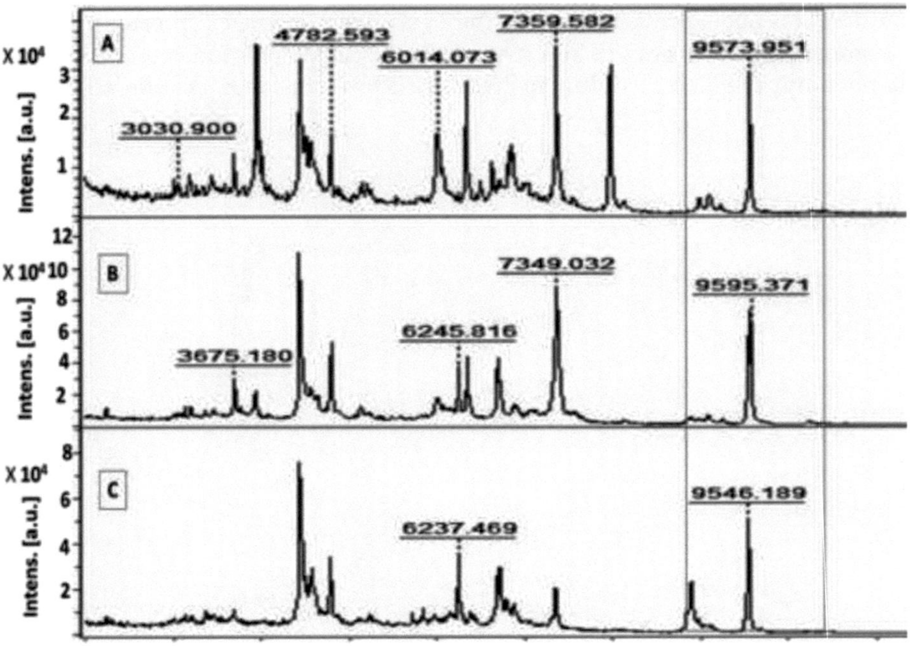

Figure 2.
Mass spectrogram of the three Enterococcus *species identified by MALDI-TOF MS analysis.* Enterococcus hirae *(A),* Enterococcus faecium *(B) and* Enterococcus fecalis *(C).*

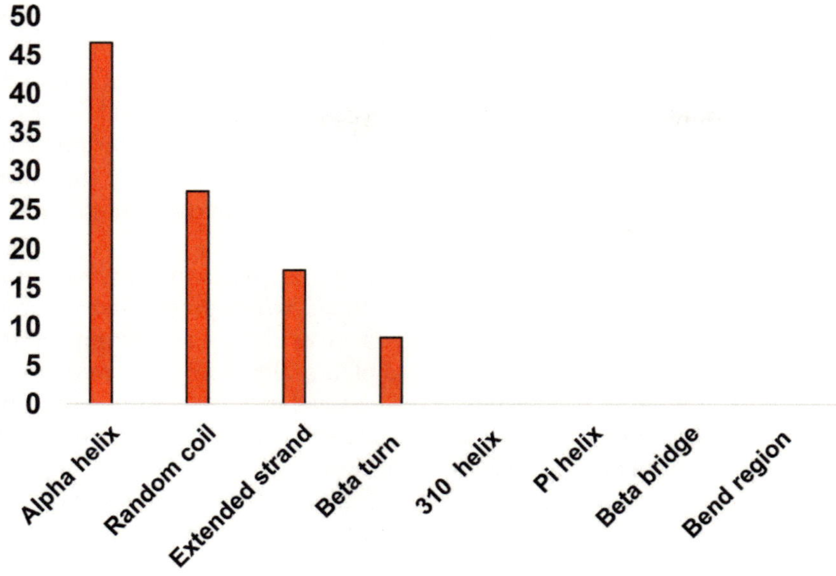

Figure 3.
Percentage of helices in Azoreductase.

Enzymes	3D-1D score (%)	ERRAT quality factor	QMEAN Z-score	AA in FR of Ramamchandran plot (%)
Azoreductase (2HPV)	99.88	95.40	1.32	97.8

Table 1.
Evaluation of structural quality of the enzyme Azoreductase (2HPV).

formation of interconnection networks was evidenced, possibly due to the interaction with bacteria that is genetically similar, which is especially favorable in dye degradation processes using bacterial consortia. At an industrial level, this improves the discoloration rate of effluents contaminated with dyes, as it has greater resistance to abiotic conditions and lower rates of enzyme inactivation, especially in large-scale operations [38].

4. Conclusions

The Bocagrande beach in Cartagena, Colombia is one of the most visited Colombian destinations by locals, as well as national and international tourists. Its high number of users throughout the year, the dumping of domestic waste generated by tourist activity, as well as other drainage carried by rain, are all considered triggers of pollution in this ecosystem.

Matrix-assisted laser ionization mass spectrometry or MALDI TOF identified other species apart from *Enterococcus faecalis*, such as *E. faecium* and *E. hirae*. The presence of these microorganisms in tourist beaches generate health related concerns about the presence of fecal contamination, sewage drains from homes or hotels along the beach, or possible overflow of wastewater treatment plants [39].

According to the World Health Organization guideline values for recreational marine waters at risk of transmitting gastrointestinal diseases (EGI) and acute febrile respiratory disease (ERFA), the results of this study indicate that Bocagrande's beaches are in category A; This means that the concentration of

Enterococcus faecalis is less than or equal to 40 CFU/100 mL and that the estimated risk for exposure is <1% for EGI and < 0.3% for ERFA.

The current biotechnological challenges lead to the development of solutions that guarantee the quality of our ecosystems and the health of human beings exposed to environmental imbalance. In relation to the problems associated with the use of dyes in different industrial processes, there have been many technological strategies developed to reduce the polluting load in industrial effluents and in receiving water bodies.

Dye removal strategies have evolved over the years. This happened due to the development of new physical and chemical methods, which progressed towards the use of environmentally friendly and cost effective biological solutions for the industry. These biological solutions have used plants, algae and other microbial biomasses as an alternative for dye removal. However, bacteria are the most robust microorganisms that, due to their structure and genome, become potential degraders of recalcitrant contaminants such as azo dyes.

The competitive advantages of bacteria are, among others, their short life cycle, their ability to adapt, and their metabolic action; they are able to degrade and detoxify the secondary metabolites produced in the discoloration process. These properties prevail in bacterial communities present in marine ecosystems, considering that these are capable of removing, in monoculture or in consortium, individual colorants, mixtures of colorants and the metal-colorant complex. Their use, although underexploited, becomes relevant with the advent of emerging technologies connected with nanotechnology, alternative energy, circular economy and environmental sustainability.

The mechanisms involved in the simultaneous removal of dyes and the metal-dye complex, the enzyme profile and the intermediate metabolites should be the subject of future studies based on genomics and proteomics. Likewise, due to the legal and environmental limitations when monitoring industrial discharges and the distribution of azo dyes in the environment, it is necessary for the scientific community to provide innovative mechanisms in which monitoring discharges and bodies of water receptors are based on amine detection.

The results of this study suggest that the enzymes Azoreductase (2HPV) are potential degraders of azo dyes due to their stability, good quality of crystallographic structure, as they are intracellular, hydrophilic and thermostable. The high content of α helices indicates their thermal resistance, which, associated with their structural quality, makes them potential degraders of azo dyes.

The properties of Azoreductase (2HPV), whose origin is *Enterococcus faecalis*, confirm that the bacterial communities present in marine ecosystems have developed special mechanisms that allow them to resist adverse environmental conditions such as hypersalinity, pH variations and the presence of heavy metals. This makes them more stable and able to degrade recalcitrant contaminants such as azo dyes [40].

Acknowledgements

This work would not have been possible without the help of the work team of the Clinical and Environmental Microbiology Group of the University of Cartagena.

Conflict of interest

The authors declare no conflict of interest.

Author details

Ganiveth María Manjarrez Paba* and Rosa Baldiris Ávila
Clinical and Environmental Microbiology Group, Faculty of Natural and Exact Sciences, University of Cartagena, Cartagena, Colombia

*Address all correspondence to: gmanjarrezp@unicartagena.edu.co

References

[1] SHAMRAIZ U, HUSSAIN RA, BADSHAH A, RAZA B, SABA S. FUNCTIONAL METAL SULFIDES AND SELENIDES FOR THE REMOVAL OF HAZARDOUS DYES FROM WATER. J. PHOTOCHEM. PHOTOBIOL. 2016; 159: 33-41. DOI: 10.1016/j.jphotobiol.2016.03.013

[2] GÜRSES A, AÇIKYILDIZ M, GÜNEŞ K, GÜRSES MS. CLASSIFICATION OF DYE AND PIGMENTS. IN DYES PIGM. SPRINGER, CHAM. 2016; 31-45. DOI: 10.1007/978-3-319-33892-7_2

[3] SINGH RL, SINGH PK, SINGH RP. ENZYMATIC DECOLORIZATION AND DEGRADATION OF AZO DYES–A REVIEW. INT. BIODETERIOR. BIODEGRAD. 2015; 104: 21-31. DOI: https://doi.org/10.1016/j.ibiod.2015.04.027

[4] BENKHAYA, S, M'RABET S, EL HARFI A. CLASSIFICATIONS, PROPERTIES, RECENT SYNTHESIS AND APPLICATIONS OF AZO DYES. HELIYON. 2020; 6:1. DOI: 10.1016/j.heliyon.2020.e03271

[5] VUTHIGANOND N, NAKPATHOM M, MONGKHOLRATTANASIT R. METAL-FREE DYEING OF COTTON FABRIC USING MANGROVE BARK POLYPHENOLS VIA AZOIC DYEING. FIBERS POLYM. 2018; 19: 2524-2532. DOI:10.1007/s12221-018-8341-6

[6] KURADE MB, WAGHMODE TR, KHANDARE RV, JEON BH, GOVINDWAR, SP. BIODEGRADATION AND DETOXIFICATION OF TEXTILE DYE DISPERSE RED 54 BY BREVIBACILLUS LATEROSPORUS AND DETERMINATION OF ITS METABOLIC FATE. JOURNAL OF BIOSCIENCE AND BIOENGINEERING. 2016; 121: 442-449. DOI: 10.1016/j.jbiosc.2015.08.014

[7] KHAN S, MALIK A. TOXICITY EVALUATION OF TEXTILE EFFLUENTS AND ROLE OF NATIVE SOIL BACTERIUM IN BIODEGRADATION OF A TEXTILE DYE. ENVIRON. SCI. POLLUT. RES. INT. 2018; 25: 4446-4458. DOI: 10.1007/s11356-017-0783-7

[8] LE, N, NUNES S. MATERIALS AND MEMBRANE TECHNOLOGIES FOR WATER AND ENERGY SUSTAINABILITY. SUSTAINABLE MATERIALS AND TECHNOLOGIES. 2016; 7: 1-28. DOI: 10.1016/j.susmat.2016.02.001

[9] THU WP, SINWAT N, BITRUS A, ANGKITTITRAKUL S, PRATHAN R, CHUANCHUEN R. PREVALENCE, ANTIMICROBIAL RESISTANCE, VIRULENCE GENE, AND CLASS 1 INTEGRONS OF ENTEROCOCCUS FAECIUM AND ENTEROCOCCUS FAECALIS FROM PIGS, PORK AND HUMANS IN THAI-LAOS BORDER PROVINCES. JOURNAL OF GLOBAL ANTIMICROBIAL RESISTANCE. 2019; 18: 130-138. DOI: 10.1016/j.jgar.2019.05.032

[10] GAO W, HOWDEN B, STINEAR T. EVOLUTION OF VIRULENCE IN ENTEROCOCCUS FAECIUM, A HOSPITAL-ADAPTED OPPORTUNISTIC PATHOGEN. CURRENT OPINION IN MICROBIOLOGY.2018; 41; 76-82. DOI: 10.1016/j.mib.2017.11.030

[11] FEDERIGI I, VERANI M, DONZELLI G, CIONI L, CARDUCCI A. THE APPLICATION OF QUANTITATIVE MICROBIAL RISK ASSESSMENT TO NATURAL RECREATIONAL WATERS: A REVIEW. MAR. POLLUT. BULL. 2019; 144: 334-350. DOI: 10.1016/j.marpolbul.2019.04.073

[12] KIM N, KIM H, PARK S, SEO G, CHO T, YU H, RHEE M. VIRULENCE PATTERNS AND PREVALENCE OF

SEVEN ENTEROCOCCUS SPECIES ISOLATED FROM MEATS AND LEAFY VEGETABLES IN SOUTH KOREA. FOOD CONTROL. 2020; 108: 106867. DOI: 10.1016/j.foodcont.2019.106867

[13] GUPTA V, KHAMPARIA S, TYAGI I, JASPAL D, MALVIYA A. DECOLORIZATION OF MIXTURE OF DYES: A CRITICAL REVIEW. GLOBAL JOURNAL OF ENVIRONMENTAL SCIENCE AND MANAGEMENT. 2015; 1: 71-94. DOI:10.7508/gjesm.2015.01.007

[14] ONG C, LEE K, CHANG Y. BIODEGRADATION OF MONO AZO DYE-REACTIVE ORANGE 16 BY ACCLIMATIZING BIOMASS SYSTEMS UNDER AN INTEGRATED ANOXIC-AEROBIC REACT SEQUENCING BATCH MOVING BED BIOFILM REACTOR. JOURNAL OF WATER PROCESS ENGINEERING.2020; 36: 101268. DOI: 10.1016/j.jwpe.2020.101268

[15] KHLIFI R, BELBAHRI L, WOODWARD S. DECOLOURIZATION AND DETOXIFICATION OF TEXTILE INDUSTRY WASTEWATER BY THE LACCASE-MEDIATOR SYSTEM. J HAZARD MATER.2010; 175: 802-8. DOI: 10.1016/j.jhazmat.2009.10.079

[16] BALAPURE K, JAIN K, CHATTARAJ S, BHATT N, MADAMWAR D. COMETABOLIC DEGRADATION OF DIAZO DYE-REACTIVE BLUE 160 BY ENRICHED MIXED CULTURES BDN. J. HAZARD. MATER. 2014; 279: 85-95 DOI: 10.1016/j.jhazmat.2014.06.057

[17] SAHASRABUDHE M, SARATALE R, SARATALE G, PATHADE G. DECOLORIZATION AND DETOXIFICATION OF SULFONATED TOXIC DIAZO DYE CI DIRECT RED 81 BY ENTEROCOCCUS FAECALIS YZ 66. JOURNAL OF ENVIRONMENTAL HEALTH SCIENCE AND ENGINEERING.2014; ,12: 151. DOI: 10.1186/s40201-014-0151-1

[18] CHAN G, RASHID N, CHUA L, NASIRI R, IKUBAR M. COMMUNAL MICROAEROPHILIC – AEROBIC BIODEGRADATION OF AMARANTH BY NOVEL NAR-2 BACTERIAL CONSORTIUM. BIORESOUR. TECHNOL. 2012; 105: 48-59. DOI: 10.1016/j.biortech.2011.11.094

[19] LIU Z, CHEN H, SHAW N, HOPPER S, CHEN L, CHEN S, WANG, B. CRYSTAL STRUCTURE OF AN AEROBIC FMN-DEPENDENT AZOREDUCTASE (AZOA) FROM ENTEROCOCCUS FAECALIS. ARCH. BIOCHEM. BIOPHYS.2007; 463: 68-77. DOI: 10.1016/j.abb.2007.03.003

[20] MAHMOOD S, KHALID A, ARSHAD M, MAHMOOD T, CROWLEY D. DETOXIFICATION OF AZO DYES BY BACTERIAL OXIDOREDUCTASE ENZYMES. CRIT. REV. BIOTECHNOL. 2016; 36: 639-651. DOI: 10.3109/07388551.2015.1004518

[21] CINPERI N, OZTURK E, YIGIT N, KITIS M. TREATMENT OF WOOLEN TEXTILE WASTEWATER USING MEMBRANE BIOREACTOR, NANOFILTRATION AND REVERSE OSMOSIS FOR REUSE IN PRODUCTION PROCESSES. J. CLEANER PROD. 2019; 223: 837-848. DOI: 10.1016/j.jclepro.2019.03.166

[22] SONI R, BHATT N, MODI H, ACHARYA P. DECOLORIZATION, DEGRADATION AND SUBSEQUENT TOXICITY ASSESSMENT OF REACTIVE RED 35 BY ENTEROCOCCUS GALLINARUM. CURRENT BIOTECHNOLOGY. 2016; 5: 325-336. DOI: 10.2174/2211550105666151211195703

[23] HALTTUNEN T, SALMINEN S, TAHVONEN R. RAPID REMOVAL OF LEAD AND CADMIUM FROM WATER BY SPECIFIC LACTIC ACID BACTERIA. INTERNATIONAL JOURNAL OF FOOD MICROBIOLOGY.2007; 114: 30-35. DOI: 10.1016/j.ijfoodmicro.2006.10.040

[24] SARAVANAN C, RAJESH R, KAVIARASAN T, MUTHUKUMAR K, KAVITAKE D, SHETTY P. SYNTHESIS OF SILVER NANOPARTICLES USING BACTERIAL EXOPOLYSACCHARIDE AND ITS APPLICATION FOR DEGRADATION OF AZO-DYES. BIOTECHNOL. REP. 2017; 15: 33-40. DOI: 10.1016/j.btre.2017.02.006

[25] HUËT MAL, PUCHOOA D. BIOREMEDIATION OF HEAVY METALS FROM AQUATIC ENVIRONMENT THROUGH MICROBIAL PROCESSES: A POTENTIAL ROLE FOR PROBIOTICS. JOURNAL OF APPLIED BIOLOGY & BIOTECHNOLOGY.2017; 5: 14-23. DOI: 10.7324/JABB.2017.50603

[26] BOTERO C, PEREIRA C, TOSIC M, MANJARREZ G. DESIGN OF AN INDEX FOR MONITORING THE ENVIRONMENTAL QUALITY OF TOURIST BEACHES FROM A HOLISTIC APPROACH. OCEAN & COASTAL MANAGEMENT.2015; 108: 65-73. DOI: 10.1016/j.ocecoaman.2014.07.017

[27] SLANETZ L, BARTLEY C. NUMBERS OF ENTEROCOCCI IN WATER, SEWAGE, AND FECES DETERMINED BY THE MEMBRANE FILTER TECHNIQUE WITH AN IMPROVED MEDIUM. JOURNAL OF BACTERIOLOGY.1957; 74:591. DOI: 13480997

[28] ZHANG L, PAN J, LIU L, SONG K, WANG Q. COMBINED PHYSICAL AND CHEMICAL ACTIVATION OF SLUDGE-BASED ADSORBENT ENHANCES CR (VI) REMOVAL FROM WASTEWATER. J. CLEANER PROD. 2019; 238: 117904. DOI: 10.1016/j.jclepro.2019.117904

[29] KASSIM A, PFLÜGER V, PREMJI Z, DAUBENBERGER C, REVATHI G. COMPARISON OF BIOMARKER-BASED MATRIX ASSISTED LASER DESORPTION IONIZATION-TIME OF FLIGHT MASS SPECTROMETRY (MALDI-TOF MS) AND CONVENTIONAL METHODS IN THE IDENTIFICATION OF CLINICALLY RELEVANT BACTERIA AND YEAST. BMC MICROBIOLOGY.207; 17: 128. DOI: 10.1186/s12866-017-1037-z

[30] HOSSEN M, HUSSAIN M, HAKIM A, ISLAM K, UDDIN M, AZAD A. BIODEGRADATION OF REACTIVE TEXTILE DYE NOVACRON SUPER BLACK G BY FREE CELLS OF NEWLY ISOLATED ALCALIGENES FAECALIS AZ26 AND BACILLUS SPP OBTAINED FROM TEXTILE EFFLUENTS. HELIYON.2019, 5: E02068. DOI: doi.org/10.1016/j.heliyon.2019.e02068

[31] WEISKERGER C, BRANDÃO J, AHMED W, ASLAN A, AVOLIO L, BADGLEY, B, JORDAO L. IMPACTS OF A CHANGING EARTH ON MICROBIAL DYNAMICS AND HUMAN HEALTH RISKS IN THE CONTINUUM BETWEEN BEACH WATER AND SAND. WATER RESEARCH.2019; 162: 456-470.DOI: 10.1016/j.watres.2019.07.006

[32] DI DATO M, GALEŠIĆ M, ŠIMUNDIĆ P, ANDRIČEVIĆ R. A NOVEL SCREENING TOOL FOR THE HEALTH RISK IN RECREATIONAL WATERS NEAR ESTUARY: THE CARRYING CAPACITY INDICATOR. SCIENCE OF THE TOTAL ENVIRONMENT. 2019; 694: 133584. DOI: 10.1016/j.scitotenv.2019.133584

[33] SOARES R, ROSSATO A, SAMBRANO G, TOLFO N, CAIERÃO J, PAIM T, D'AZEVEDO P. EVALUATION OF A SELECTIVE CHROMOGENIC MEDIUM FOR DETECTING VANCOMYCIN-RESISTANT ENTEROCOCCI. BRAZILIAN JOURNAL OF MICROBIOLOGY.2017; 48: 782-784. DOI: 10.1016/j.bjm.2017.03.005

[34] ANGARITA M, FORERO E, GUTIÉRREZ N, YAÑEZ F, ROMERO

C. ANALYSIS OF *ENTEROCOCCUS FAECALIS*, STAPHYLOCOCCUS AUREUS, AND CANDIDA ALBICANS IN CAST METAL CORES. JOURNAL OF THE FACULTY OF DENTISTRY, UNIVERSITY OF ANTIOQUIA. 2017; 28: 292-310.DOI: 10.17533/udea.rfo.v28n2a4

[35] SOHN K, PECK K, JOO E, HA Y, KANG C, CHUNG D, SONG J. DURATION OF COLONIZATION AND RISK FACTORS FOR PROLONGED CARRIAGE OF VANCOMYCIN-RESISTANT ENTEROCOCCI AFTER DISCHARGE FROM THE HOSPITAL. INTERNATIONAL JOURNAL OF INFECTIOUS DISEASES.2013; 17: 240-246. DOI: 10.1016/j.ijid.2012.09.019

[36] TURKOVICOVA L, SMIDAK R, JUNG G, TURNA J, LUBEC G, ARADSKA J. PROTEOMIC ANALYSIS OF THE TERC INTERACTOME: NOVEL LINKS TO TELLURITE RESISTANCE AND PATHOGENICITY. JOURNAL OF PROTEOMICS.2016; 136: 167-173.DOI: 10.1016/j.jprot.2016.01.003

[37] SARKAR S, BANERJEE A, CHAKRABORTY N, SOREN K, CHAKRABORTY P, BANDOPADHYAY R. STRUCTURAL-FUNCTIONAL ANALYSES OF TEXTILE DYE DEGRADING AZOREDUCTASE, LACCASE AND PEROXIDASE: A COMPARATIVE IN SILICO STUDY. ELECTRON. J. BIOTECHNOL.2020; 43: 48-54. DOI: 10.1016/j.ejbt.2019.12.004

[38] UNNIKRISHNAN S, KHAN M, RAMALINGAM K. DYE-TOLERANT MARINE ACINETOBACTER BAUMANNII-MEDIATED BIODEGRADATION OF REACTIVE RED. WATER SCIENCE AND ENGINEERING. 2018; 11(4), 265-275. 10.1016/j.wse.2018.08.001

[39] CONLEY K, CLUM A, DEEPE J, LANE H, BECKINGHAM B. WASTEWATER TREATMENT PLANTS AS A SOURCE OF MICROPLASTICS TO AN URBAN ESTUARY: REMOVAL EFFICIENCIES AND LOADING PER CAPITA OVER ONE YEAR. WATER RESEARCH X. 2019; 3: 100030.DOI: 10.1016/j.wroa.2019.100030

[40] ZHUANG M, SANGANYADO E, ZHANG X, XU L, ZHU J, LIU W, SONG H. AZO DYE DEGRADING BACTERIA TOLERANT TO EXTREME CONDITIONS INHABIT NEARSHORE ECOSYSTEMS: OPTIMIZATION AND DEGRADATION PATHWAYS. J. ENVIRON. MANAGE.2020. 261: 110222. DOI: 10.1016/j.jenvman.2020.110222

Chapter 6

Mangrove in Ecuador: Conservation and Management Strategies

Fausto López-Rodríguez

Abstract

In Ecuador, 100% of the mangroves are protected through different mechanisms: protected areas, community mangrove concessions, and protective forests. However, there is still deforestation of the mangroves, even in protected areas, which is caused mainly by the construction/expansion of shrimp pools. Shrimp is currently Ecuador's first non-oil export product. The Sustainable Use and Mangrove Custody Agreements are very important because they cover an area almost similar to that of protected areas. This mechanism is effective because it allows the sustainable extraction of resources from the mangrove, but forces the "custodians" to protect this ecosystem. This chapter includes a case study on the management of the "mangrove concessions" of the province of El Oro, southern Ecuador, in which the management effectiveness of these areas is analyzed. We found that despite the limited resources that these mangrove concessions have, the level of management is "satisfactory", which means that most of the management objectives are met. However, these areas should receive more support, both from the state and private organizations, as they conserve more than 40% of Ecuador's mangrove.

Keywords: mangrove, mangrove concessions, local communities, sustainable, management, protected areas, shrimp pools

1. Introduction

Mangrove forests have a great relevance for the world, they are resilient to the adverse effects exerted by both natural and anthropic factors and play a fundamental role in the strategy for adaptation to climate change. They are also the basis for food security for ancestral communities and traditional users.

At a global level, it is calculated that mangrove forests cover an area of about 15′000,000 hectares, 11% (≈1′650,000 hectares) of which is found on the Pacific and Caribbean coasts of South America. In Ecuador, there are an estimated 161,000 hectares of mangroves. Regarding mangrove losses, annual values of 0.16% have been estimated worldwide between 2000 and 2012, with South America having the lowest rates of deforestation compared to Asia, Africa, North and Central America [1]. Ecuador had the highest deforestation in the 1970s with 27% deforestation going from 300,000 ha in the 1960s to 145,000 ha in the 1980s. The mangrove conservation efforts in Ecuador were based, originally, on the issuance of policies and laws and the creation of state protected areas [2]. A new mechanism was created in 1999, which recognized the rights and traditional uses of the communities that lived in these ecosystems or that depended on their resources for their survival: The Custody and Sustainable Use of Mangrove Agreements (AUSCM).

This conservation mechanism plays a key role in the conservation of the mangrove in Ecuador, since it occupies almost the same extension as the mangrove that is found in the protected areas of the National System of Protected Areas (SNAP). It complements government conservation efforts, with community conservation activities.

The main objectives of this study are to show the key role of the AUSCM in the conservation of the mangrove in Ecuador, since it occupies almost the same extension as the mangrove that is found in the protected areas of the National System of Protected Areas (SNAP) as well as the effective management that local communities give to this ecosystem.

2. Current situation of mangroves in Ecuador

Ecuador has a land area of 256,370 km^2 and an estimated population, as of 2018, of 17′096,789 inhabitants. It limits to the north with Colombia, to the south and east with Peru and to the west with the Pacific Ocean. In addition, on the maritime border of the Galapagos Islands, it borders the Cocos Island of Costa Rica [3]. The Ecuadorian territory includes, in addition to the land area, 1′092,140 km^2 of maritime area, which is equivalent to 4.3 times the continental territory. The extension of the continental coast is 1200 km [4] (**Figure 1**).

In the coastal of Ecuador, several economic, social and environmental activities are developed, with diverse interests that cause that the Coastal region is a highly conflictive region. This region has several marine-coastal ecosystems, one of the most important is the mangrove. This ecosystem is considered one of the most productive in the world, however, it is one of the most threatened. In the case of Ecuador, in the span of 40 years, 27% of the mangrove were lost as a consequence, mainly, of the construction of shrimp pools and urban expansion [2].

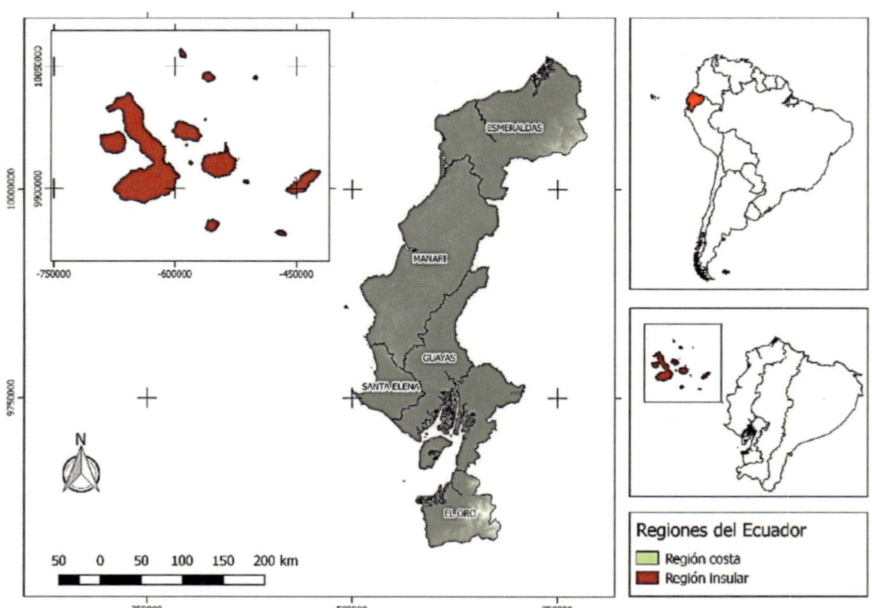

Figure 1.
Coastal and insular region of Ecuador.

The mangrove swamp in Ecuador is located mainly in six estuaries. CLIRSEN and PMRC (2007) determined that in 1969 the mangrove area in Ecuador reached 203,624 hectares, which, by 2006, was reduced to 146,971 hectares, a loss of 56,653 hectares, equivalent to 22,8% [2].

The highest annual rate of deforestation occurred between 1991 and 1995 (2.35% per year); on the contrary, between 1995 and 1999 a recovery of the mangrove cover was observed and between 1999 and 2006, the annual rate of deforestation was 0.13% [2]. For its part, in the decade from 2006 to 2016, a mangrove recovery of 14,864 hectares was observed again [2, 5] (**Table 1**).

The current mangrove area in Ecuador is 161,835 ha. This mangrove coverage gain was the result of reforestation programs implemented by different institutions and communities that have mangrove use and custody agreements, as well as the natural regeneration of the ecosystem [6, 7].

At the policy and legal level, these marine-coastal ecosystems have received significant support. In the first place, there is the 2007 Constitution of Ecuador that recognizes mangroves as "fragile and threatened ecosystems" which gives them a special "status" in relation to other types of ecosystems. The Organic Environment Law (2017) confirmed earlier regulations that, in Ecuador, all mangrove forests are property of the State. Another important regulation, based on the economic valuation of ecosystem services of mangroves, is the resolution No. 056 of the Ministry of the Environment of 2011, based on which fines for mangrove deforestation of up to USD 89,273 are established [8]. According to a study carried out by the Charles Darwin Foundation (CDF), Galapagos National Park Directorate (GNPD) and the Scripps Institute of Oceanography at the University of San Diego, each hectare of mangrove in Galapagos has at least a value of $ 27.852, because the 3,700 hectares store more than 778.000 tons of carbon. Its conservation is considered a measure of adaptation and mitigation to climate change.

On April 4, 2019, the National Action Plan for the Conservation of Mangroves in Ecuador was approved. This Plan seeks to promote the protection, recovery and sustainable use of mangroves, with a focus on improving the quality of life of ancestral and traditional users [9].

Estuaries	1969	1984	1987	1991	1995	1999	2006	% of mangrove lost comparted to1969
Cayapas Mataje	23.677	23.653	23.507	22.863	21.947	22.057	21.400	9.6
Muisne	3.282	2.701	2.445	1.340	830	1.187	1.187	52.5
Cojimíes	13.123	9.917	8.466	6.028	3.651	1.921	2.742	79.1
Chone	3.973	1.673	1.040	784	391	705	932	76.5
Guayaquils Gulf	124.320	119.277	115.784	109.608	102.108	104.715	105.130	15.4
Jambelí Archipelago	34.712	24.592	23.570	21.092	17.697	19.111	19.111	56.2
Total	**203.624**	**181.815**	**174.815**	**161.718**	**146.628**	**149.699**	**146.971**	**27.6**

Source: [2].

Table 1.
Mangrove loss by estuary (in hectares). Period 1969–2006.

3. Causes of mangrove loss in Ecuador

Anthropogenic activities are the main causes of the destruction of mangroves. These activities substantially alter the composition, structure and function of mangroves, reducing the ecosystem services they provide. In Ecuador, the transformation of the mangrove into shrimp pools and urban development are the main factors in the loss of this ecosystem.

3.1 Shrimp activity

Ecuador is currently one of the main producers of farmed shrimp in the world. Although the shrimp industry is of great importance to the country, it has also been the main cause of the destruction of the mangroves.

According to the National Chamber of Aquaculture of Ecuador, in 2015 there were 213,000 hectares assigned to shrimp production, of which 181,000 hectares are located in an area that was originally mangrove. The province of Guayas has the highest coverage of shrimp farms, with approximately 140,000 hectares, which represents 66% of the total production followed by the province of El Oro with 18% [10]. In addition to the deforestation of mangroves that the construction of the pools implies, during the activity itself, effluents rich in organic and inorganic particles are released that deteriorate the resources of the estuaries [11].

Shrimp is currently Ecuador's first non-oil export product. The Ministry responsible for the fishery sector estimated that, in 2013, 66% (6,192 ha) of the shrimp ponds in the province of Esmeraldas were illegal, in the province of El Oro 39% (12,576 hectares), and Manabí 59% (8,434 hectares) and Guayas 18% (17,437 hectares) [12]. Even protected areas have not been exempted from the presence of this activity since in 4 out of 6 coastal protected areas there were shrimp farms installed or expanded after the creation of the area.

CLIRSEN and PMRC (2007) determined that the main loss of mangrove coverage occurred between 1969 and 1995, a period in which more than 56,000 hectares were deforested. Since 2006, mangrove coverage has remained relatively stable while the area covered by shrimp farms increased to 210,000 hectares [2]. The evolution of areas dedicates to shrimp farming is showed in **Table 2**.

The loss of mangroves up to 2000 occurred at the same time that shrimp farming was developing. A slight increase in the mangrove area is observed as of 2000, which allows us to deduce that the conservation strategies and the regulations created generated favorable impacts for the conservation of this ecosystem (**Figure 2**).

	Year								
	1969	**1984**	**1987**	**1991**	**1995**	**1999**	**2000**	**2006**	**2016***
Shrimp farms (ha)	0	89,368	117,728	145,998	178,071	175,253	175,253	175,748	210,000
Mangroves (ha)	203,695	182,157	175,157	162,186	146,938	149,556	127,690	148,230	161,835

**2018 data do not show a significant increase in the area of shrimp farms.*
Source: [2].

Table 2.
Evolution of areas dedicated to shrimp farming and mangrove conservation in the 1984–2016 period.

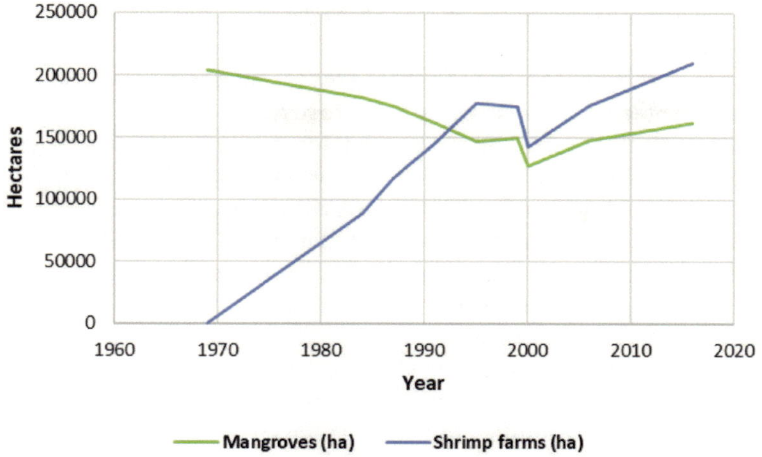

Figure 2.
Relationship between mangrove loss and shrimp farm area growth.

3.2 Urban development

Many coastal cities are located in areas that were once occupied by mangroves. It is estimated that until 1994 approximately 3,000 to 5,000 hectares of mangroves were destroyed to make way for the growth of cities such as Guayaquil, Machala, and Esmeraldas [13].

4. Strategies for mangrove conservation in Ecuador

The accelerated transformation of the mangroves of continental Ecuador led to the generation and implementation of different strategies to protect them, thereby achieving that, at present, 100% of the mangrove is protected. The first measures focused on establishing laws and regulations to protect the mangrove. In 1986, the entire mangrove was classified as a protective forest, providing governmental institutions with a basic set of instruments to punish its deforestation. Later, in 1994, a mangrove deforestation ban was established, and the expansion and construction of new shrimp ponds was prohibited.

The main conservation strategies are:

a. Mangroves declared as Protective forests

b. Mangroves declared Protected areas

c. Sustainable use and Custody Agreements of mangrove forests

d. Governmental incentive program for forest conservation: Socio Manglar

4.1 Protective forests

Al mangroves were declared as Protective forests in 2003, is a figure that in addition to conservation, allows the development of certain activities like deforestation, contamination etc. [14].

4.2 National system of protected areas (SNAP)

The first formal strategy to stop the rapid loss of mangroves in Ecuador was declaration of protected areas covering mangrove forests. In 1979 the Churute Mangrove Ecological Reserve was established, located in the Guayas province and 16 years later the following was established: Cayapas Mataje Ecological Reserve (Esmeraldas). The most recent protected area with mangrove ecosystem, the Area Nacional de Recreación Isla Santay, was created in 2010 [15]. Today, Ecuador has 19 marine and coastal protected areas, which together represent about 8% of the total coverage of the SNAP. Nine of these areas contain (totally or partially) 72,523.48 hectares of mangrove [15] (**Table 3**).

In 2017 the Network of Marine and Coastal Protected Areas of Ecuador (Red de AMCPs) was created, as a mechanism for political-administrative interaction to enhance institutional resources and manage the areas in an articulated and synergistic manner. Coastal marine protected areas. The aim of this network is to guarantee biological connectivity between ecosystems by creating connectivity corridors and conserving the biodiversity of the National System of Protected Areas in the marine-coastal zone [16].

However, even though the laws forbid deforestation and degradation of protected areas, the mangrove coverage reduced even in the most recent year (2010 to 2018) by 150.34 hectares due to conversion to shrimp farms, which shows a weakness in the control, surveillance and monitoring of these areas and an illegality on the part of the shrimp farms [17].

4.3 The mangrove sustainable use and custody agreements (AUSCM)

In Ecuador, all mangrove forests are property of the State, with the Ministry of the Environment (MAE) being the institution responsible for their management [18].

The AUSCM are the management tool contemplated in the Ecuadorian legal framework [18, 19], under which mangrove forests are handed over to ancestral users to custody these areas. The AUSCM guarantee the "custodians" exclusive access to the mangrove areas with the right to sustainably use bio-aquatic resources,

	Protected area	Month/Year it was created protected area	Total Extention (ha)
1	Reserva Ecológica Manglares Churute	Julio 1979	49,389
2	Reserva Ecológica Cayapas Mataje	Octubre 1995	51,300
3	Reserva Ecológica Arenillas	Mayo 2001	13,170
4	Refugio de Vida Silvestre Islas Corazón y Fragatas	Octubre 2002	2,811
5	Reserva de Producción de Fauna Manglares El Salado	Noviembre 2002	10,635
6	Refugio de Vida Silvestre Manglares Estuario del Río Muisne	Marzo 2003	3,173
7	Refugio de Vida Silvestre Manglares El Morro	Septiembre 2007	10,030
8	Refugio de Vida Silvestre Estuario del Río Esmeraldas	Junio 2008	242
9	Área Nacional de Recreación Isla Santay	Febrero 2010	2,215

Source: http://areasprotegidas.ambiente.gob.ec.

Table 3.
State protected areas with partial or total mangrove coverage.

but in turn have the obligation to realize control and surveillance of mangrove and report the progress of its management to the [19]. These agreements, also called "mangrove concessions," are important because they protect 42.85% of the Ecuadorian mangrove (almost the same extent as the protected areas) and are the livelihood for thousands of families living in the coast. In the province of El Oro, the AUSCM cover about 80% of this ecosystem [7].

Currently in Ecuador there are 59 community organizations with custody agreements with an area of 69,369 hectares (**Figure 3**).

The fact that the mangrove area under AUSCM is almost equal to that occupied by the protected areas of the National System of Protected Areas (SNAP) is relevant (**Table 4**).

4.4 Socio Manglar, incentive for mangrove conservation

The Socio Bosque Program has a special incentive for mangrove protection named "Socio Manglar", which provides an economic incentive to "custodies" (communes, associations, etc.) of mangrove forests. The mangrove conservation incentive was created in 2014 to support the management of the Sustainable Use and Custody of Mangrove Forests Agreements (AUSCM). As of 2020, there are 27 signed agreements covering 34,160 hectares, for which they receive USD 413,481 each year, with which 1,635 families benefit. In the province of El Oro, the 11

Figure 3.
Evolution of the AUSCM since 2000.

Distribución del manglar	Cobertura (ha)	%
Protected Areas (SNAP) (2020)	72.523	45.15
Sustainable Use and Custody of Mangrove Forests Agreements (2020)	69.369,48	42.86
Protection Forest (2020)	19.943	11.99
TOTAL	**161.835**	**100**

Source: www.ambiente.gob.ec.

Table 4.
Mangrove distribution according to conservation status.

beneficiary organizations of Socio Manglar receive approximately USD 113,092.54 for the conservation of 5,343.18 hectares of mangroves [20].

5. The AUSCM in the province of El Oro

The mangrove ecosystem in the El Oro province (southern Ecuador) covers 19,318 hectares that represent 4% of the provincial surface. The area covered with mangrove forest is located on the coastal zone and the Jambelí Archipelago in which there are six small fishermen villages, which depend on the resources they extract from the mangrove, particularly the collection of black shell (*Anadara tuberculosa* and *A. similis*) and red crab (*Ucides occidentalis*), and artisanal fishing [21].

It is estimated that originally the mangrove coverage in the El Oro province occupied 35,144 hectares, having been reduced, by 2006, to 16,152 hectares, which is equivalent to a loss of 56.2% [2]. According to the Ministry of Environment and Water, there is a recovery of 2,866 hectares between 2006 and 2018. **Figure 4** shows the evolution of mangrove coverage in the El Oro Province.

There are currently 24 AUSCM in the province of El Oro that comprise 15,666.34 hectares, which means that 81% of the mangrove swamp is protected in these areas. The Arenillas Ecological Reserve, with an area of 13,170 hectares, has only 1,239 hectares of mangroves, which makes the role of the AUSCM in the conservation of the mangroves in this province even more relevant. These custody areas benefit 1,323 families.

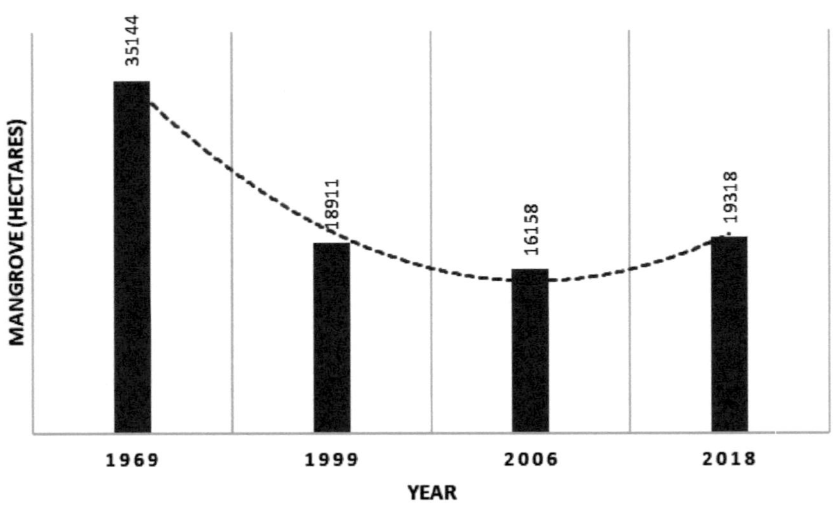

Figure 4.
Evolution of mangrove coverage in the El Oro Province. Period 1969–2018. Source: [2].

6. Main drivers of mangrove degradation in the El Oro province

6.1 Population growth

The constant growth of the population of the El Oro province has been the consequence of a series of factors, mainly associated with the development of economic activities, among which are: the banana industry, shrimp farming and mining. Between 1974 and 2018, the population almost tripled, currently reaching a population of around 700,000 inhabitants. Only five of the 14 cantons in the province of El Oro have mangroves, but 413,299 people live there, representing 68.8% of the total population. 49% of these inhabitants reside in the Machala canton, where Puerto Bolívar is located, the main axis of development of this province. Urban and industrial development has caused the deforestation of mangrove forests in the El Oro Province [7].

6.2 Aquaculture development policy (shrimp farming)

The development of shrimp farming in Ecuador began in 1970. It had a rapid growth due to the issuance of a series of favorable public policies and the endowment of economic resources from international organizations such as the World Bank, the International Monetary Fund and the Inter-American Development Bank [22]. The felling of the mangrove for the construction of the shrimp ponds is the most important impact, but during the shrimp cultivation process other environmental impacts are generated, such as discharges (biocides, fertilizers, antibiotics, etc.) that are released into the estuaries and sea without any treatment [7, 10]. The province of El Oro is the second with the largest extension of shrimp farms in the country, with an estimated area of 35,576.6 hectares, which represents 19.05% of its territory. The shrimp concessions are superimposed on the mangrove that is under Sustainable Use and Custody Agreements, in 333.3 hectares, that is, 2.3% of the area with AUSCM [21].

6.3 Agricultural development policy (banana and cocoa)

Bananas was, for decades, Ecuador's first non-oil export product, which is why it has received important support from the State, through subsidies, tax benefits and other incentives designed to consolidate its production and export. In the province of El Oro, banana plantations cover more than 58,000 hectares, which represents more than two-thirds of the permanent crop area of the province (68%). This monoculture uses various chemicals that pollutes the waters that go to the estuaries [23].

6.4 Mining development policy

Mining activity has received a great boost, particularly in the last 12 years, during which the legal and political framework was reformed to promote mining activity, especially large-scale mining. The approval of a new Mining Law and its respective regulations stand out here. In the province of El Oro there are 853 mining concessions, which cover 154,785 hectares, equivalent to 28.4% of its surface. In 11 of the 14 cantons of this province there are concession areas. The problem is that most of the mining concessions are located in the upper part of the watershed basins polluting rivers that later deposit their contaminated waters in estuaries and mangroves [7].

6.5 Climate change and El Niño phenomenon

Another important driver at the global scale is the El Niño phenomenon that is increasing due to climate change. The "El Niño Phenomenon" (ENSO-El Niño-Southern Oscillation) is a climatic event that occurs approximately every 2–7 years and is related to short-term climate variability. It results in the appearance of warmer (El Niño) or colder (La Niña) surface waters than normal in the central and eastern tropical Pacific [24]. It has been determined that among the most important changes caused by El Niño are the increase in sea temperature and sea level that can produce more intense rains than normal. The decrease in salinity due to the contribution of fresh water from the rains, can interfere in the development and growth of the species *Rhizophora mangle* and *R. harrisonni* (they need and are tolerant to high levels of salinity) and in that of mollusks and crustaceans that depend on the habitat of these species, particularly the black or brown shell (*Anadara tuberculosa* and *A. similis*) and the red crab (*Ucides occidentalis*) [25].

7. Evaluation of the management of the AUSCM

Evaluating the effectiveness of the management of protected areas or other conservation measures is key because it allows, in addition to knowing the management problems and their causes, to identify and apply, in a timely manner, strategies and measures to improve their management. There are several methodologies to evaluate the management of protected areas or other conservation measures. In Ecuador, the principal method used to evaluate the effectiveness of management conservation measures is the Hockings Reference Framework proposed by the IUCN (2000) and the 360° performance evaluation [26].

In the case of the 20 AUSCM in the province of El Oro in force in 2017, the methodology known as 360° Performance Evaluation was applied with some adaptations which allows verifying compliance with the inherent obligations of the organizations involved, as well as the factors that affect their actions [27]. The 17 indicators that were used are divided into four groups:

a. Current state of the mangrove. Evaluates aspects such as mangrove coverage and pollution.

b. Compliance with the agreement. It includes compliance with the Management Plan, the delivery of semi-annual reports and complaints, etc.

c. Custodian performance. It includes compliance with the implementation of control and surveillance programs, sustainable use, participation of fishermen, economic contributions, commercialization, etc.

d. Performance of support entities. It includes the MAAE and other institutions of the national and local government, academy, etc.

For the rating and weighting of this evaluation, the Likert scale was used with four rating levels (from 0 to 4) associated with a percentage that reflects the respective management levels. This method is based on a method used by De Faría (1993) and later incorporated by WWF, GIZ and IUCN in the Manual for Evaluating the Management Effectiveness of Protected Areas [28, 29]. Ulloa et al. (2012) applied it in the evaluation of the management effectiveness of five coastal marine protected areas [30]. **Table 5** shows the levels of qualification of management effectiveness of the areas under custody.

The results indicated that the management effectiveness of the 20 mangrove custody areas analyzed is in ranges between 46.7% and 93.5%, which means that none of these areas has an unsatisfactory management (**Figure 5**).

At the individual indicator level, the highest corresponds to the maintenance and increase of the mangrove coverage and the lowest to "Direct sale/added value of crabs and shells " which shows that most of the fishermen still work with intermediaries who are in charge of commercialize bioaquatic products [26].

The perceptions of the custodians regarding to the recovery of the mangrove were verified through a multi-temporal analysis in three periods: 1985–1999-2018

Level I. Unsatisfactory. (≤ 25%). Indicates that the protected area has not guaranteed its long-term permanence
Level II. Slightly Satisfactory. (26–50%). Means that the protected area is highly vulnerable to conjunctural factors and its permanence is not guaranteed in the long term
Level III. Satisfactory. (51–75%). Indicates that the protected area has deficiencies which prevent an effective management, but the management objectives are partially met
Level IV. Very Satisfactory. (76–100%). Indicates that the permanence of the protected area is guaranteed and management objectives are fully meet.
Source: [28, 30].

Table 5.
Management effectiveness levels of areas under AUSCM.

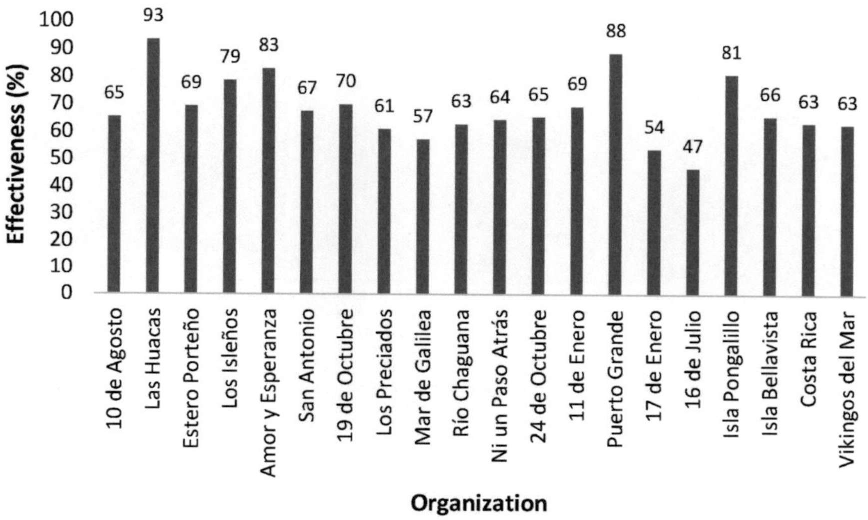

Figure 5.
Management effectiveness of the 20 mangrove custody [26].

on the Jambelí Archipelago. The multi-temporal analysis agrees with the perception of organizations and institutions that mention that the massive felling of mangroves has stopped since the creation of the custody agreements (AUSCM).

The mangrove cover in the Jambelí archipelago in 1985 was 13,084.17 hectares, while in 1999 it was reduced to 8,978.43 hectares. For 2018 the mangrove cover was 9,454.47 hectares. While between 1985 and 1999 in the Jambelí archipelago 4,106 hectares of mangroves were deforested, between 1999 and 2018, 476 hectares were recovered precisely at the time when mangrove concessions began to be awarded to local communities (**Figure 6**).

There are small clearings, mainly in the areas adjacent to the shrimp ponds, for the maintenance of its walls. On the other hand, they perceive mangrove recovery in sites far from shrimp farms, which they attribute to the control and surveillance and reforestation activities carried out by the organizations [31].

The Socio Manglar incentive has contributed to improve the management effectiveness of the mangrove custody agreements, as well as the organizational strengthening. However, the overexploitation of bioaquatic resources and contamination still persist, requiring greater support and inter-institutional coordination from control entities such as the Ministry of Environment and Water of Ecuador (MAAE), Ministry of Production, Foreign Trade, Investments and Fisheries (MPCEIP) and the Public Prosecutor's Office [26].

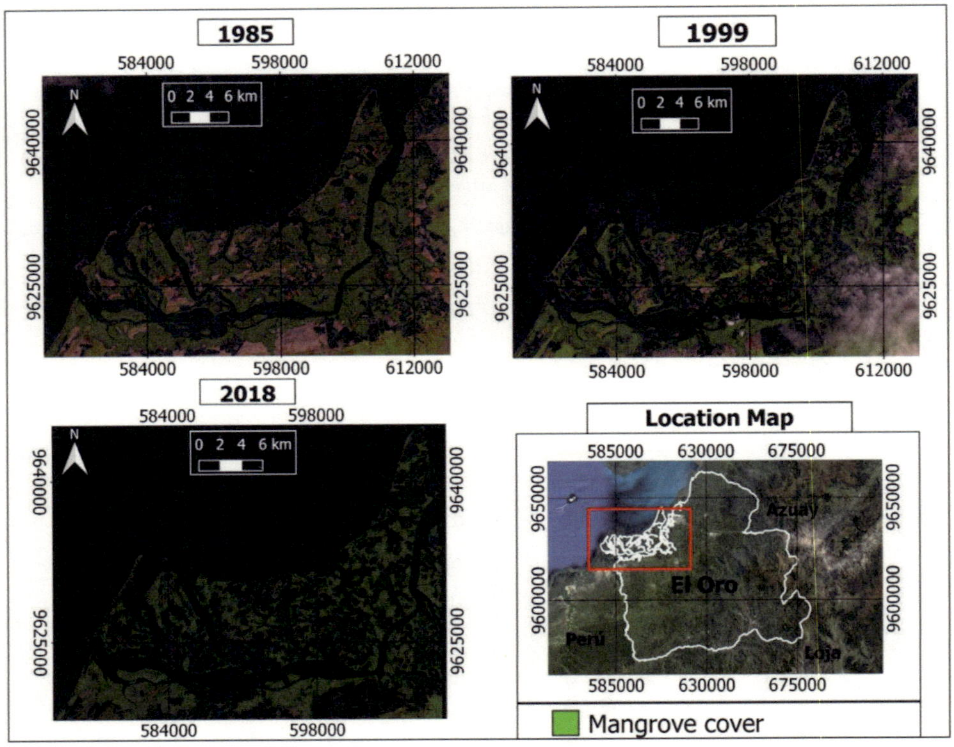

Figure 6.
Mangrove cover on Jambelí Archipelago in three periods: 1985–1999-2018.

8. Key elements in the management of community areas

Several key elements have been developed to strengthen the management of the AUSCM in the province of El Oro. Among the most prominent are:

Capacity building program for beneficiaries of the AUSCM

This training included the topic of organizational, administrative-financial strengthening and management of marine-coastal resources.

Capacity building program for Technical Advisors of the AUSCM

It is a series of 6 modules aimed both for partners of organizations in charge of technical aspects and for organizations that provide.

ManglarApp

The ManglarApp mobile application is a digital tool that is part of the global trend of electronic government and digital citizenship, and that allows complaints, notifications and early alerts of anomalies in mangroves through a smartphone. This application was created in order to improve communication between the partners of organizations with AUSCM and the control entities in matters of mangroves like the Ministry of Environment and Water of Ecuador[1], Ministry of Production, Foreign Trade, Investments and Fisheries and Prosecutor's Office) [32].

[1] On March 4, 2020, through executive decree 1007, President of Ecuador ordered the merger of the Ministry of the Environment (MAE) and the Secretariat of Water (SENAGUA) creating the Ministry of Environment and Water. With this change, the Undersecretary of Marine and Coastal Management, an entity that was decentralized, disappeared, and now operates from Quito (the capital of Ecuador). The functions now fulfilled by the Under Secretariat of Natural Heritage.

9. Conclusions and recommendations

Few evaluations of management effectiveness have been performed in the AUSCM. The last evaluation was made in 2008 [33]. One of the main differences in the results of this study was the mangrove cover. While the 2008 assessment reports a loss of mangroves in the custody areas, the present investigation found a recovery of this ecosystem. Likewise, this research found greater compliance in the execution of management plans than the 2008 study. Another important difference is found in the Socio Manglar incentive, which did not exist in 2008 which is an important support for the management of the AUSCM.

The declaration of the mangroves of Ecuador by the Organic Environment Law (2017) as a national asset is an important milestone in mangrove ecosystem conservation because it gives this ecosystem a high-level legal protection, which is complemented by the recognition given by the Constitution of Ecuador as a "fragile and threatened" ecosystem.

Although all mangroves in Ecuador are protected through different mechanisms (protected areas, AUSCM and protective forests), it has been identified that, even within the areas protected by the State, the loss of mangroves continues, as shrimp farms convert mangrove forests into shrimp pools.

Both in the protected areas of the SNAP and in the areas under AUSCM, a land occupation by shrimp farming is observed, which shows that the monitoring systems of these areas is lacking behind, hence the convenience of strengthening monitoring systems through the use of satellite images, as well as other community monitoring mechanisms is evident.

Shrimp farming continues to be a factor in the loss of mangroves both in the State's protected areas (SNAP) and in the custody areas.

The elimination of the Under Secretariat of Marine and Coastal Management could imply a weakness in the support of the Ministry of the Environment and Water to the management of the AUSCM.

Organizations with custody areas, government and technical assistance entities, agree that Sustainable Use and Custody Agreements are an effective tool for the conservation and for the economy of ancestral communities and traditional users of the mangrove. It has been effective in curbing mangrove logging (as demonstrated by the multitemporal study), although overexploitation of bioaquatic resources and contamination of water and sediment persists, threats for which greater support and inter-institutional coordination are required by the authorities. Control entities (MAE, MPCEIP, Prosecutor's Office).

Mangrove Custody and Sustainable Use Agreements are an important and effective conservation strategy for the conservation of mangroves in the province of El Oro. The recovery of mangrove coverage and effectiveness of the mangrove management is "Satisfactory".

The Socio Manglar program has supported the organizations, providing financial means to improve their control and surveillance, pay basic organizational costs (office, material), however not all organizations receive this incentive, so it is necessary to seek new sources of financing to improve the success of the AUSCM without the Sociomanglar incentive.

Author details

Fausto López-Rodríguez
Departamento de Ciencias Biológicas, Research Group: Gobernanza, Biodiversidad y Áreas Protegidas, Universidad Técnica Particular de Loja, Loja, Ecuador

*Address all correspondence to: fvlopezx@utpl.edu.ec

References

[1] Giri, C, E. Ochieng, L.; L. Tieszen, Z., Zhu, A., Singh, T., Loveland, J. Masek and N. Duke. (2010). Status and distribution of mangrove forests of the world using earth observation satellite datageb_584 154.159.

[2] CLIRSEN-Centro de Levantamientos Integrados de Recursos Naturales por Sensores Remotos y PMRC-Programa de Manejo de Recursos Costeros. (2007). Actualización del estudio multitemporal de manglares, camaroneras y salinas en la costa ecuatoriana al 2006.

[3] SETEMAR-Secretaría Técnica del Mar. (2014). Políticas Públicas Costeras y Oceánicas: Diagnóstico y propuesta de implementación. Biótica Cía. Ltda. Eds. Guayaquil: Editorial El Telégrafo.

[4] SENPLADES (Secretaria Nacional de Planificación y Desarrollo). (2017). Plan de Ordenamiento del Espacio Marino Costero-POEMC.

[5] Ministerio del Ambiente del Ecuador. (2017). Regularización de camaroneras. Consultado de: http://www.ambiente.gob.ec/proyecto-regularizacion-de-camaroneras/.

[6] Ministerio del Ambiente del Ecuador. (2017). Guía de derechos y deberes de las organizaciones custodios del manglar. Ministerio del Ambiente de Ecuador, Conservación Internacional Ecuador, Instituto Humanista para la Cooperación con los Países en Desarrollo, Organizaciones de las Naciones Unidas para la Alimentación y la Agricultura y Fondo para el Medio Ambiente Mundial. Guayaquil, Ecuador.

[7] López-Rodríguez, FV. (2018). Mangrove Concessions: An Innovative Strategy for Community Mangrove Conservation in Ecuador. In: Makowski C and Finkl CW (eds) Threats to Mangrove Forests: Hazards, Vulnerability, and Management. Cham: Springer International Publishing, pp. 557-578. DOI: 10.1007/978-3-319-73016-5_25.

[8] Ministerio del Ambiente del Ecuador. (2011). Resolución 056. Registro Oficial No. 496 de 21 de Julio de 2011.

[9] Carvajal R. y X. Santillán. (2019). Plan de Acción Nacional para la Conservación de los Manglares del Ecuador Continental. Ministerio del Ambiente de Ecuador, Conservación Internacional Ecuador, Organización de las Naciones Unidas para la Educación, la Ciencia y la Cultura (UNESCO) y la Comisión Permanente del Pacífico Sur (CPPS). Proyecto Conservación de Manglar en el Pacífico Este Tropical. Guayaquil, Ecuador.

[10] Cámara Nacional de Acuacultura. (2017). Análisis de las Exportaciones de Camarón Diciembre – 2016. Consultado de: http://www.cna-ecuador.com/

[11] Hurtado, M. y Rodríguez, T. (2012). Caracterización de los ecosistemas marinos y su conectividad. Presentado en el Taller Ecosistemas Marinos y su Conectividad MAE-GIZ, Manta.

[12] Gobierno de la República del Ecuador (2008) Acuerdo Ministerial 149, RO No 412, del 27 de agosto de 2008. Ministerio de Agricultura, Ganadería, Acuacultura y Pesca.

[13] Grupo Spurrier (2012) Estudio del impacto de la acuicultura camaronera en el Ecuador. Guayaquil.

[14] Presidencia de la República del Ecuador. (2003). Texto Unificado de Legislación Secundaria de Medio Ambiente. Decreto Ejecutivo 3516. Registro Oficial Edición Especial 2 de 31-mar. 2003.

[15] Fundación Futuro Latinoamericano, "Gobernanza en las Áreas Protegidas

Marinas y Costeras: el caso del Ecuador", Quito, 2011, 40 p.

[16] Ministerio del Ambiente del Ecuador. (2017). Crea la Red de Áreas Marinas y Costeras Protegidas. Registro Oficial 77 de 12-sep.-2017

[17] Castro, R. (2019). Determinación de la cobertura vegetal/uso actual del suelo y las dinámicas de cambio (2010- 2018) en las áreas protegidas marino costeras mediante la utilización de imágenes satelitales. Molina Moreira, N. & Galvis, F. (Comp). Primer Congreso Manglares de América. Universidad Espíritu Santo. Samborondón-Ecuador.

[18] Código Orgánico del Ambiente-CODA. (2017). Registro Oficial Nro. 983 de 12 de abril de 2017. Asamblea Nacional República del Ecuador.

[19] Ministerio del Ambiente del Ecuador. (2010). Procedimiento de Acuerdos de Uso Sustentable y Custodia de Manglar. Acuerdo Ministerial 129. Registro Oficial 283, 21 de septiembre de 2010.

[20] Ministerio del Ambiente y Agua. (2020). Tabla de convenios Socio Manglar. Documento no publicado.

[21] Universidad Técnica Particular de Loja (UTPL) y Deutsche Gesellschaft fuer Internationale Zusammenarbeit (GIZ) GmbH. (2018). Hacia un manejo adaptativo de los ecosistemas costeros de la provincia de El Oro, Ecuador. Sistematización de la aplicación de la metodología Manejo Adaptativo de Riesgo y Vulnerabilidad en Sitios de Conservación (MARISCO). Quito – Ecuador. UTPL (*C. Naranjo*, F. López, M. Morocho, E. Toledo y M. Riofrio.) y GIZ

[22] Romero, N. (2014). Neoliberalismo e industria camaronera en Ecuador. Letras Verdes. Revista Latinoamericana de Estudios Socioambientales N. o 15, marzo 2014. FLACSO

[23] León, L. (2017). La sostenibilidad ambiental en el sector productivo bananero del cantón Machala. Universidad Técnica de Machala. II Congreso Internacional de Ciencia y Tecnología.

[24] INOCAR. (2019). El Niño: Generalidades. Consultado de: https://www.inocar.mil.ec/modelamiento/elnino/nino_generalidades.php

[25] Morera, S., Condom, T., Crave, C., Steer, P. y Guyot, J., (2017). The impact of extreme El Niño events on modern sediment transport along the western Peruvian Andes (1968-2012). Scientific Reports.

[26] López-Rodríguez, A, Benítez, I, Jurrius. (2019). Efectividad de Manejo de Acuerdos de Uso Sustentable y Custodia de Manglar en la provincia de El Oro. Martha Molina Moreira (Comp.) Primer Congreso Manglares de América, Guayaquil, Ecuador

[27] Coello, S., D. Vinueza & R. Alemán. (2008). Evaluación del desempeño de los acuerdos de uso sustentable y custodia de manglar de la zona costera del Ecuador. Ministerio del Ambiente del Ecuador – Conservación Internacional – Unión Mundial para la Naturaleza (UICN) – Comisión Mundial de Áreas Protegidas de UICN – Programa de apoyo a la gestión descentralizada de los recursos naturales en las tres provincias del norte del Ecuador (PRODERENA) – Ecobiotec. Julio de 2008: 52pp.

[28] De Faria, H. (1993). Elaboración de un procedimiento para medir la efectividad de manejo de áreas silvestres protegidas y su aplicación en dos áreas protegidas de Costa Rica, Turrialba, Costa Rica, Tesis. Mag. Sc., CATIE, 1993.

[29] Cifuentes, M. & A. Izurieta. (1999). Evaluation of protected area management effectiveness: analysis of

procedures and outline for a manual. WWF Centroamérica. Turrialba, Costa Rica.

[30] Ulloa, R. y Tamayo, D. (2012). Evaluación de efectividad de manejo de cinco áreas protegidas marinas y costeras del Ecuador continental: Parque Nacional Machalilla, Reserva Marina Galera-San Francisco, Refugio de Vida Silvestre Manglares El Morro, Refugio de Vida Silvestre Marino Costero Pacoche y Reserva de Producción Faunística Marino Costero Puntilla de Santa Elena. Ministerio del Ambiente del Ecuador y Conservación Internacional Ecuador. Guayaquil, Ecuador.

[31] Universidad Técnica Particular de Loja (UTPL). (2017). Evaluación de efectividad de manejo de los acuerdos de uso sustentable y custodia del manglar en la provincia de El Oro (Documento digital). SGMC, CI-Ecuador, HIVOS, GEF y FAO. Loja. Ecuador.

[32] Reyes-Bueno, F, Jurrius, I, López-Rodríguez, F, Astudillo, D. y Ramírez-Moreina, L. (2019). ManglarApp: una herramienta tecnológica de gobierno electrónico que facilita la comunicación entre usuarios del manglar y entes de Control sobre las amenazas socio-ambientales en los manglares. Molina Moreira, N. & Galvis, F. (Comp). Primer Congreso Manglares de América. Universidad Espíritu Santo. Samborondón-Ecuador.

[33] Coello, S., D. Vinueza & R. Alemán (2008). Evaluación del desempeño de los acuerdos de uso sustentable y custodia de manglar de la zona costera del Ecuador. Ministerio del Ambiente del Ecuador – Conservación Internacional – Unión Mundial para la Naturaleza (UICN) – Comisión Mundial de Áreas Protegidas de UICN – Programa de apoyo a la gestión descentralizada de los recursos naturales en las tres provincias del norte del Ecuador (PRODERENA) – Ecobiotec. Julio de 2008: 52 pp. + 4 Figuras +17 Tablas +5 Apéndices +29 mapas.

Section 5

Coastal Geodynamics

Chapter 7

Current Geodynamics and Evolutionary Trends of a Headland Bay Beach System in the Semi-Arid Coast of Chile

María-Victoria Soto, Misael Cabello and Joselyn Arriagada-González

Abstract

The Chilean coast is controlled by the tectonics and structure, generating an irregular coastal landscape, with bays, marine terraces, sandy and gravel beaches, sand dune fields and Andean slopes, forming some mega cliffs that are attacked by waves. The Chilean coastline is shaped by headland bay beaches, with a dynamic coast modeled by south-western winds and south–north longshore current. We analyzed the case of the Coquimbo mega headland bay beach, which consists of four headland bay beaches. A methodological study was carried out on the morphometric parameters of the shoreline and the types of beaches dominated by waves along with geomorphological analysis of the coastal zone. We observed a mass transfer process from south to north. The northern sections of the bays are the places with the densest sand dune fields. This concentration of dunes occurs in each bay individually and in the mega bay as well. The sedimentary supply comes from Andean catchments to the shoreline and is transported and reworked by the longshore current to the northern area, where a huge sand field dune has developed, 120 km away from the mouth of Limarí River, the most southern catchment in the study area. In the mega bay, the current trend is a continuous sedimentary supply, despite the semi-arid conditions and the extreme drought that has affected the area since 2011. The study area is also a popular destination in Chile for beach tourism and is a place of interest for the mining industry.

Keywords: headland bay beaches, sand dunes, Andean catchments, sandy supplies

1. Introduction

The first approach to headland bay beaches, logarithmic bays and crenulated bays was realized by Halligan (1906 in Ref. [1]), [2–7], and recently by [8–11]. The concept of bays as units for analysis in coastline territories and the theories related to headland bay beaches are particularly important for the comprehension of the geodynamic processes of the Chilean coastline [12], due to the geographical position of the country in the subduction zones of the South-American and Nazca plates, involving a dynamic tectonic movement along the Chilean and western South-America coastal configuration.

IntechOpen

In this context, and from a morpho-structural perspective, we were able to explain the configuration of irregular coastline associated to tectonics and structural controls in the study area [13–15]. We established the influence of relevant factors in the geometry of the Chilean coastline, what the morphometric conditions of the coastline are, the longshore current, the angle of waves incidence and the types of beaches dominated by waves. These variables were then used to create a morphologic model and a process-response system [16–20]. The impact on the headland bay beaches shows a systematic distribution of wave energy in the longshore current direction. The headland bay beaches are a complex system of mass transfer and form evolutions which are controlled by the structure, the tectonics and the lithology of the area [21, 22].

The study area is Coquimbo Bay, which is located between the mega headland Punta Lengua de Vaca and the sand dunes of Los Choros to the extreme north of the bay. This zone is a headland bay beach system (**Figure 1**). These systems have not been studied much in Chile, the few studies that have been carried out include [1, 17, 18, 21–25].

Figure 1 shows the general geographical context of the study area. The bays feature the river mouths of the Andean catchment which are the areas that supply mass to the littoral. In the high Andean catchment, some remnant glaciers still remain, with a glacial-snowy-pluvial system, which generates a permanent flow to the Limari and Elqui rivers, despite the intense drought that has been affecting the area since 2010–2011. The catchments (in green) are mainly coastal and subject to rainfall patterns, as well as the low Andean Coastal Range.

In the study area, the big headland of Punta Lengua de Vaca in the north, which is 7.5 km long, plays a role in protecting the bay from SW winds, which show a morpho-sedimentological expression as far away as the big sand dune field, 120 km further north. Between both limits of the Coquimbo mega bay, marine terraces, paleo dunes, beach ridges and sandy beaches are found.

It has been demonstrated that in Chile, sand dunes are concentrated in the northern area of bays, due to the effect of the headlands in wave refraction associated with the prevailing SW–NE winds and the longshore current from the river mouths to the bays [12, 17–19, 21, 23, 25–27].

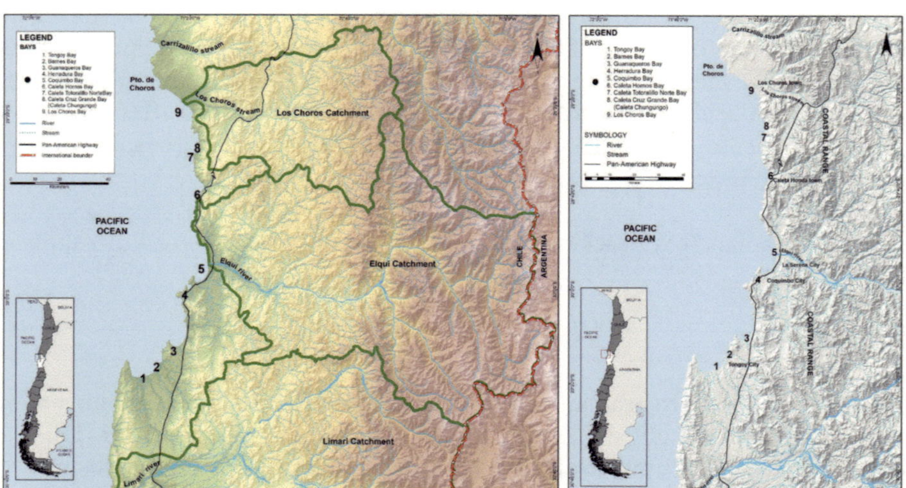

Figure 1.
Geographical context of bays and catchment in the Coquimbo mega bay, Chile. Details of the research environment: Headland bay beach system from Punta Lengua de Vaca headland to Los Choros sandy beach with huge sand dune fields.

2. General geomorphology of Coquimbo mega bay

The geomorphological map of the area of study (**Figure 2**) shows the different groups of identified forms, associated with its location on the western side of the coastal range. The geomorphological map has been created by utilizing existing information [28–34] and fieldwork data.

The hillslopes of the Coastal Range have been formed by intrusive and volcanic rocks and are in direct contact with the coast. These bays have marine terraces, Pleistocene and Holocene sand dunes, beaches and platforms of active abrasion, reefs, cliffs and small bays of rocks or gravel. These characteristics are very important in the northern area of the mega bay where Cabello [35] identified up to three levels of marine abrasion platforms. These sectors are mostly uninhabited but they are subject to a great deal of economic and environmental interest such as mining extraction activities, protected natural areas such as marine parks and artisanal fishing creeks used by indigenous people (Changos people). These different uses of the land and the sea are not compatible with each other; hence, the area is subject to latent environmental conflicts.

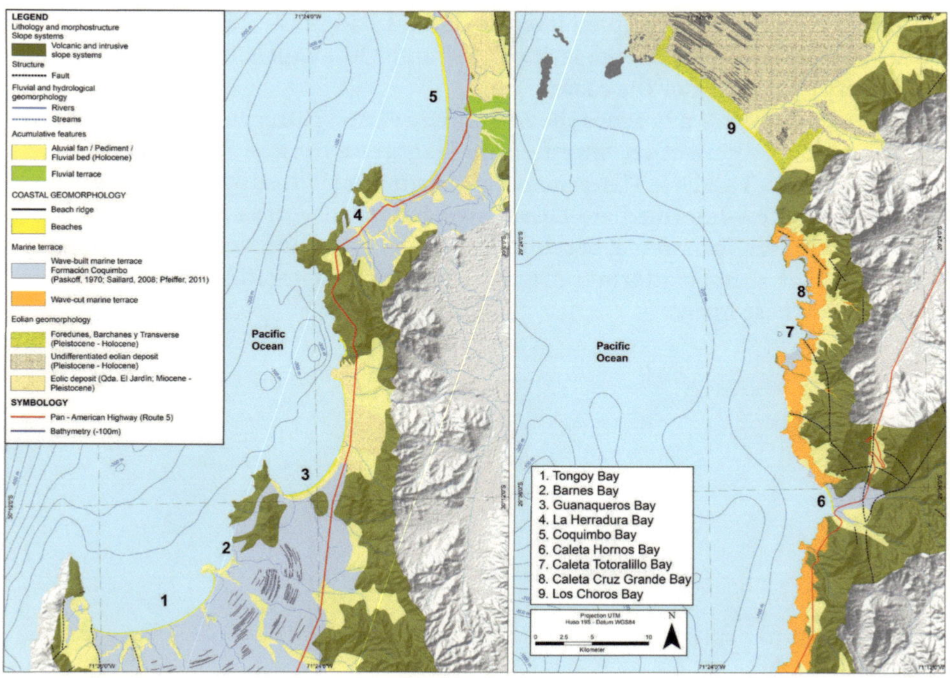

Figure 2.
Geomorphological map of the study area.

3. Littoral zone

The analysis of the littoral zone was achieved by using the classification of wave-dominated beaches by Wright & Short in Ref. [36], complemented with the morphometry of the shoreline by [1, 22].

The mid and southern zones of the area feature bays and sandy beaches. Tongoy bay (N°1 in **Figure 2**) in the southern part of the Coquimbo mega bay has been formed by the effect of the great headland named Punta Lengua de Vaca, which is 7.5 km long. Tongoy is a Reflective beach with low energy, with breaking waves smaller than one-meter-height; this Reflective pattern was observed during the 2007–2019 period. The structural influence in the bay can be quantified through the values of the coastal area morphometrics, with a 67° asymmetry angle and a 95° refraction angle (**Figures 3** and **4**).

The neighboring bay, Guanaqueros (N° 3 in **Figure 2**) is a 17 km long sandy beach. The wave dominated-beach type varies between Reflective with low energy, in the south and Intermediate (Longshore Bar through Rhythmic Bar and Beach and Transverse Bar and Rip) in the north. It has an asymmetry angle of 357° in the northern part and 44° in the southern part. Due to the local headland presence, the refraction angle reaches 20° (**Figures 3** and **4**).

Coquimbo Bay has a wide coastline strip of approximately 15 kilometers of sandy beach. This bay is protected by a rocky point which forms Coquimbo's peninsula in the southern part and is an obstacle to the prevailing SW winds and their associated wave action. We observed a systematic distribution of the wave energy from south to north, similar to the theoretical model, which implies a Reflective-Intermediate-Dissipative beach in the southern, center and northern sectors of the bay. From the point of view of the relative position of the shoreline, Coquimbo Bay has an asymmetry angle of 353° and a refraction angle of the surge action of 26° (**Figures 3** and **4**).

The mid-northern area is a rocky coast hosting cliffs and mixed sand and gravel beaches. It has little bays sculpted into the Coastal Range. Morphometric values are variable with asymmetry angles of 6°, 58° and 330°, which illustrate the strong irregularity of this part of the coast.

The wave dominated beaches showed widely varied patterns and were identified as low energy Reflective, Intermediate Transverse Bar and Rip, and Low Tide Terrace (**Figures 3** and **4**). The systematic distribution pattern of wave dominated beaches, Reflective-Intermediate-Dissipative, is hard to verify except for in a few little bays with sandy or gravel beaches. This pattern was not identifiable for rocky beaches with abrasion platforms or reefs because of the alteration of the surge generated by these forms.

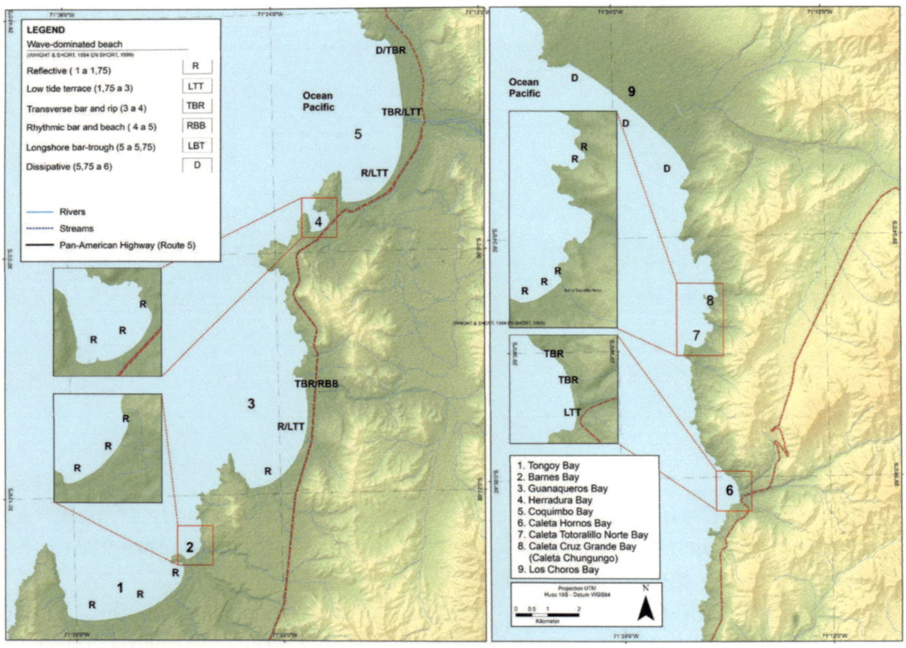

Figure 3.
Wave-dominated beach. Source: Based on [37].

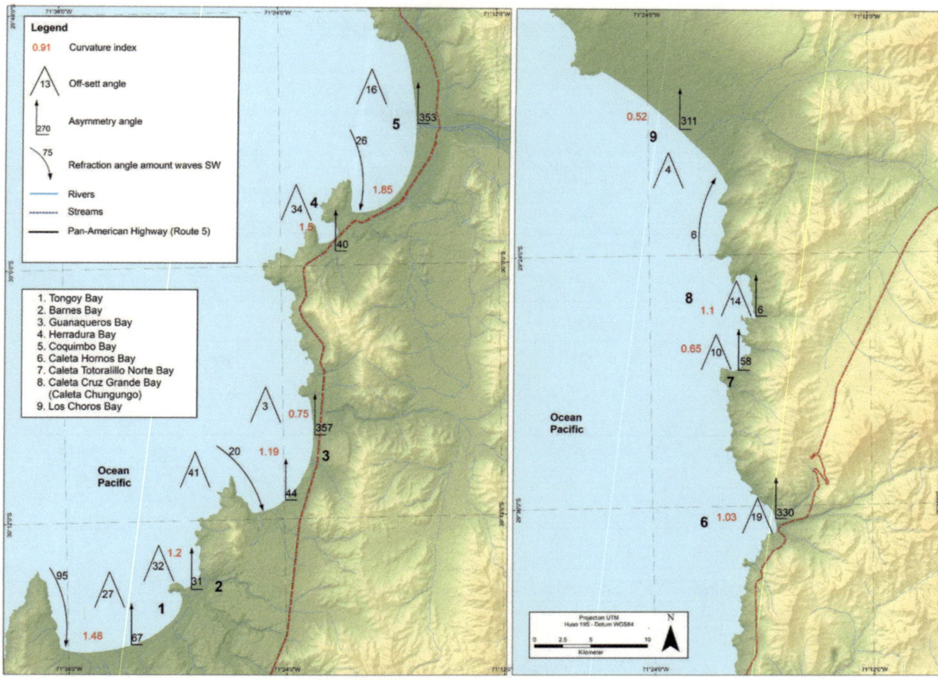

Figure 4.
Coastline morphometry of the mega bay of Coquimbo. Source: Based on [37].

In the northern area, Los Choros sand dune field is clearly distinguished. The coastline is a 16 km long sandy beach with a big dune field which is 15 km wide, which climbs the hillslope of the Coastal Range. Theses dunes have been stable since Holocene – Pleistocene [38].

The northern zone of the mega bay is the accumulation zone of the general system. It is a Dissipative beach, with breaking waves higher than 2 meters. Nonetheless, by analyzing the beach as an individual subsystem we observed some variations in the type of wave-dominated beaches, between Intermediate-Dissipative (2002, 2007, 2013 and 2014). During the years of analysis, we never observed a Reflective state. Villagrán [26] established that this zone shows a bathymetry associated to a lower relative depth, functioning as a trap for sediment entry. The relative position of the coastline (NW-SE orientation), which is transverse to the predominant wind system, shows an asymmetry angle of 311° and a refraction angle of 6° (**Figures 3** and **4**).

Table 1 shows the relationships between the offset angle, the refraction angle of the waves and the type of wave-dominated beaches for 8 bays that have been analyzed. Also, we can see how these latter form the mega bay system, with a dynamic relationship in terms of longshore current direction and associated sediment transfer. Tongoy Bay has an offset angle of 27° in the southern area and 4° in the northern area (**Figures 3** and **4**).

The dissipation of energy up to the middle and northern zones of each bay is characteristic of the presence of headlands in big bays. The southern zone, with a 95° refraction angle, shows a predominance of wave-dominated beaches of Reflective type and low energy; the middle zone with a rocky coast tends to be more Intermediate and the northern zone, with a 6° refraction angle is Dissipative with high energy, which matches the biggest deposit of dunes in the mega bay. The headland bay beach model is totally applicable to the Coquimbo mega bay. It also applies to the smaller bays that compose the mega bay's system.

N°	Bay	Asymmetry Angle	Refraction Angle	Off-set Angle	Waves-dominant Beach
1	Tongoy	67	95	27	R
2	Barnes	31	—	32	R
3	Guanaqueros	44S / 357 N	20	3	R/LTT - TBR/RBB
4	Herradura	40	—	34	R
5	Coquimbo	353	26	16	R/LTT - TBR/ LTT - D/TBR
6	Caleta Hornos	330	—	19	TBR
7	Caleta Totoralillo Norte	58	—	10	R
8	Caleta Cruz Grande (Chungungo)	6	—	14	R
9	Los Choros	311	6	4	D

Reflective (R); Low tide terrace (LTT); Transverse bar & rip (TBR); Rhythmic bar & beach (RBB); Dissipative (D)

Table 1.
Morphometric parameters in Coquimbo's structural bays system and classification of wave-dominated beaches.

4. Beach-dune relationship

Previous studies on headland bay beaches in central Chile have allowed for the establishment of the conditions for a beach-wave interaction system [12, 16–22, 24, 25, 27, 39–41], the morphologic expression of which are foredunes and transgressive dunes. This dynamic system is associated with the coastline orientation and the balancing of internal mass, the changing structure and the balancing of external mass, as well as the relationship with wave-dominated beaches (**Figures 3** and **4**). The presence of foredunes and transgressive dunes are directly related to the availability of sediments from their supplying sources (mainly from the Andean catchment) and also the capacity of transportation in the littoral zone.

By analyzing the geomorphological map (**Figure 2**), we identified that there are foredunes in each bay with sandy beaches. We found the following sequence of forms in Tongoy (**Figure 5**): Holocene vegetated foredunes-terraces with beach ridges-vegetated transverse dunes [39]. In Coquimbo (N°3 in **Figure 2**), there is a sequence of vegetated foredunes and beach ridges (destroyed by urban expansion) and Pleistocene sand dunes. In the extreme north of the mega bay, we observed the following sequence (**Figure 6**): foredunes-active transgressive dunes and the mega field of stabilized sand dunes.

Figure 5.
Beach ridge succession in the marine terrace of the Holocene. Source: Fondecyt project 1120234.

Figure 6.
Vegetated foredunes in Los Choros dissipative beach. May 2014. Source: Fondecyt project 1120234.

Figure 7.
Mixed sand and gravel beach, vegetated-foredunes eroded by the waves. Source: Fondecyt project 1120234.

In the case of the study area, the permanence of foredunes was noted. Pulse erosion and seasonal deposits existed and we identified that the foredune reconstructs itself [12, 21], thus reinforcing the conditions of a sediment transfer to sandy beaches. This is important to highlight due to the fact that in the area's semi-arid climate there has been a drought for the past decade that has reduced the volume of the Andean catchment. Nonetheless, we have seen evidence of changes in Los Choros dune system, indicating that the development of embryonic dunes and foredunes, which in conjunction with barchan and elongated dunes are evidence of the current supply of sand to the beach. Another relevant factor is the strong condition of erosion in the mixed sand.

The condition of a predominantly Dissipative wave-dominated beach is a consequence of the condition of obliquity (4° of offset angle, **Table 1**), a bathymetry of superficial platform that facilitates/provides the transfer of sediments from the south via the longshore current (**Figure 7**).

The beach-dune relationship in the system of bays that constitute the Coquimbo mega bay shows a dynamic of constant sand supply leading to a positive sediment budget. This is demonstrated each time that sandy beaches show sequences of beach-active foredunes, even if they are seasonally eroded or affected by offshore storms or tsunamis [12].

5. Final considerations in respect to the present dynamic and evolutionary trend of the Coquimbo mega bay

We analyzed the current geodynamics conditions of the Coquimbo mega bay coastal zone in terms of littoral morphology, beaches and dunes, as indicators of

Figure 8.
Tongoy beach, oblique section of the Coquimbo mega bay. Erosion of the sandy beach due to the impact of September 2015 tsunami and reconstruction of the same beach in November 2016, showing evidence of sand supply to the beach [12].

mass transfer. Previous studies have analyzed the relationship between the semi-arid river catchment and the current dynamic coastal processes, focusing mainly on the source areas for sedimentary supply, the pulse of delivery of mass to the shoreline and responses to extreme climatic events. It has been possible to identify an increase in the erosional process in the sandy beaches, nevertheless, with seasonal patterns [27, 42].

The characteristics of wind deposits in the Coquimbo mega bay show that the general distribution of the sand dynamic is replicated in each individual bay and all of them constitute the Coquimbo mega bay where the biggest concentration of sand is accumulated in the northern part of the bay (Los Choros), which corresponds to the oblique zone of the system.

From the dynamic system point of view, Los Choros dune field is very similar to Hesp's scenario model (2013 in Ref. [12]); as a matter of fact, it had a sequence of nearshore-beach-foredunes-transgressive dunes and the evolution of old and present dunes. This dynamic condition has been verified through observation over a 20-year period, showing the permanence of a sandy beach with embryonic dunes and foredunes as evidence of sediment supply.

Foredunes and embryonic dunes are also present in the Reflective low energy beaches of the oblique zone in the south. This condition has been verified as a seasonal dynamic trend. The extreme events of the 2015 tsunami (**Figure 8**), winter and offshore storms over the past.

years have generated a systematic process of destruction of the foredunes which then are newly rebuilt, proving that sediment supply comes from external sources other than the local catchment [12, 21].

The mouth of the Limari River located in the Coastal Range mega cliff does not have a dune deposit matching the size of Andean catchment. As a consequence, it shows that the sand of the Limari river supplies the beaches in the south, Coquimbo partially and mainly the dunes in the north.

Acknowledgements

Would like to thank Fondecyt for financing the Project "Geodynamics and evolutionary trend of the coastal system of the mega bay of Coquimbo: Towards a prognosis of natural hazards for scenarios of endogenous and exogenous environmental changes" (Fondecyt 1120234).

Author details

María-Victoria Soto[1]*, Misael Cabello[2] and Joselyn Arriagada-González[1]

1 Department of Geography, Universidad de Chile, Santiago, Portugal

2 Lab. of Physical Geography, Department of Geography, Universidad de Chile, Chile

*Address all correspondence to: mvsoto@uchilefau.cl

References

[1] Araya-Vergara JF. Influencias morfogenéticas de los desalineamientos y líneas de costa contrapuestas en el litoral de Chile Central. Informaciones Geográficas. 1983;23:9-29. DOI: 10.5354/0719-5370.2013.27674.

[2] Jennings JN. The influence of wave action on coastal outline in plan. The Australian Geographer. 1955;6(4):36-44. https://doi.org/10.1080/00049185508702306.

[3] Davies JL. Wave refraction and the evolution of shoreline curves. Geogr. Studies. 1959;5:1-14.

[4] Yasso W. Plan geometry of headland bay beaches. Journal of Geology. 1965;73:702-714. DOI: 10.2307/30079652.

[5] Le Blond P. An explanation of the logarithmic spiral plan shape of headland bay beaches. Journal of Sedimentary Petrology. 1979;49(4):1093-1100. https://doi.org/10.1306/212F78BA-2B24-11D7-8648000102C1865D.

[6] Hsu JRC, Evans C. Parabolic Bay Shapes and Aplications. Proceedings Institution of Civil Engineers. 1989; 87(2): 556-570. https://doi.org/10.1680/iicep.1989.3778.

[7] Short A. Handbook of beach and shoreface morphodynamics. Ed. Andrew Short, Wiley. 1999. ISBN: 978-0-471-96570-1.

[8] Rodríguez-Polo S, Del Río L, Benavente J. Longitudinal distribution of slope and sediment characteristics in headland-bay beaches in Cádiz, Spain. Journal of Coastal Research. 2018;85(10085):306-310. DOI: 10.2112/SI85-062.1.

[9] El-Shinnawy1 A, Medina1 R, González M. Equilibrium planform of headland bay beaches: Effect of directional wave climate. Coastal Dynamics. 2017;20:749-759.

[10] Li B, Zhuang Z, Cao L, Du F. Application of the Static Headland-Bay Beach Concept to a Sandy Beach: A New Elliptical Model. Journal of Ocean University of China. 2020;19:81-89. https://doi.org/10.1007/s11802-020-3899-1.

[11] Tasaduak S, Weesakul S. Experimental study on dynamic equilibrium of headland-bay beaches. Journal of Coastal Conservation. 2016;20;165-174. https://doi.org/10.1007/s11852-016-0427-y.

[12] Soto MV. Assessment of process dynamics and evolutionary trend of the western part of the arid Chilean coastal range: relationships between river catchments and coastal dynamics of the Coquimbo bay system, Chile [Thesis]. Tubingen: Eberhard Karls Universitat; 2017.

[13] Le Roux JP, Gómez C, Venegas C, Fenner J, Middleton H, Marchant M, Buchbinder B, Frassinetti D, Marquardt C, Gregory-Wodzicki KM, Lavenu A. Neogene–Quaternary coastal and offshore sedimentation in north-central Chile: Record of sea level changes and implications for andean tectonism. Journal of South American Earth Sciences. 2005;19:83-98. https://doi.org/10.1016/j.jsames.2003.11.003.

[14] Le Roux JP, Olivares D, Nielsen S, Smith N, Middleton H, Fenner J, Ishman S. Bay sedimentation as controlled by regional crustal behavior, local tectonics and eustatic sea level changes: Coquimbo formation (Miocene-Pliocene), bay of Tongoy, central Chile. Journal Sedimentary Geology. 2006;184:133-153. DOI: 10.1016/j.sedgeo.2005.09.023.

[15] Binnie A, Dunai T, Binnie S, Victor P, González G, Bolten A. Accelerated late quaternary uplift revealed by 10Be exposure dating

of marine terraces, Mejillones Peninsula, northern Chile. Quaternary Geochronology. 2016;36:12-27. DOI: 10.1016/j.quageo.2016.06.005.

[16] Araya-Vergara JF. Toward a classification of beach profiles. Journal of Coastal Research. 1986;2 (2):159-165. DOI: 10.2307/4297157.

[17] Martínez C. El efecto de ensenada en los procesos litorales de las ensenadas de Valparaíso, Algarrobo y Cartagena, Chile central [Thesis]. Santiago. Universidad de Chile; 2001.

[18] Soto MV. Aspectos morfodinámicos de ensenadas desalienadas del litoral de Chile central. Pichilemu y Caleta Los Piures. Revista de Geografía Norte Grande. 2005;33:73-87. https://repositorio.uc.cl/handle/11534/10470.

[19] Arriagada J. Geomorfología estuarial comparada en la zona semiárida de Chile. Casos de Copiapó y Choapa [Thesis]. Santiago. Universidad de Chile; 2009.

[20] Arriagada J, Soto MV, Sarricolea P. Morphodynamic Environment in a semiarid mouth river complex Choapa river, Chile. In: Marghany M, editor. Handbook of Advanced Geoscience Remote Sensing. Demand: BoD-Books; 2014; p. 254-271. http://dx.doi.org/10.5772/57410. ch11.

[21] Magallanes V. Relaciones morfodinámicas de la línea de costa entre Tongoy y las dunas de Los Choros. Transferencia sedimentaria en la mega ensenada de Coquimbo [Thesis]. Santiago: Urbanismo, Universidad de Chile; 2017.

[22] Soto MV, Arriagada J, Cabello, M. The Accretional Beach Ridge System of Tongoy Bay: An Example of a Regressive Barrier Developed in the Semiarid Region of Chile. Recent Advances in Petrochemical Science. *2018;*4(4):001-008. DOI: 10.19080/RAPSCI.2018.04.555641.

[23] Araya-Vergara JF. Sistema de interacción oleaje-playa frente a los ergs de Chanco y Arauco, Chile. Gayana Oceanológica. 1996;4(2):159-167.

[24] Soto MV, Arriagada J. Características dinámicas de ensenadas estructurales de Chile centra. Maitencillo-Cachagua y Papudo, Región de Valparaíso. Revista de Geografía Norte Grande. 2007; 38: 99-112. http://dx.doi.org/10.4067/S0718-34022007000200006.

[25] Martínez C, Quezada M, Rubio P. Historical changes in the shoreline and littoral processes on a headland bay beach in central Chile. Geomorphology. 2011;135:80-96. DOI: 10.1016/j.geomorph.2011.07.027.

[26] Villagrán C. Dinámica costera en el sistema de bahías comprendidas entre Ensenada Los Choros y Bahía Tongoy, región de Coquimbo [Thesis]. Santiago: Universidad de Chile; 2007.

[27] Soto MV, Märker M, Rodolfi G, Sepúlveda SA, Cabello M. Assessment of geomorphic processes affecting the paleo-landscape of Tongoy bay, Coquimbo region, central Chile. Geografia Fisica e Dinamica Quaternaria. 2014;37(1):51-66. DOI 10.4461/GFDQ.2014.37.6.

[28] Paskoff R. Le Chili Semi-aride, recherches geómorphologiques. Bourdeaux: Biscaye Fréres; 1970. 420 p. Traducción al español José Enrique Novoa Jerez. Ediciones Universidad de La Serena. La Serena; 1993.

[29] Ota Y, Paskoff R. Holoceno deposits on the coast of north – central Chile: radiocarbono ages and implications for coastal changes. Revista Geológica de Chile. 1993;20:25-32. DOI: http://dx.doi.org/10.5027/andgeoV20n1-a03.

[30] Saillard M. Dynamique du soulèvement côtier Pleìstocène des Andes centrales: Etude de l'evolution géomorphologique et datations (10Be)

de séquences de terrasses marines (sud Pérou – nord Chili [Thesis]). Toulouse: Université de Toulouse; 2008.

[31] Saillard M, Hall SR, Audin L, Farber DL, Hérail G, Martinod J, Regard V, Finkel RC, Bondoux F. Non-steady long-term uplift rates and Pleistocene marine terrace development along the Andean margin of Chile (31s) inferred from 10Be dating. Earth and Planetary Science Letters. 2009;277:50-63. https://doi.org/10.1016/j.epsl.2008.09.039.

[32] Saillard M, Hall SR, Audin L, Farber DL, Regard V, Hérail G. Andean coastal uplift and active tectonics in southern Peru: 10Be surface exposure dating of differentially uplifted marine terrace sequences (San Juan de Marcona, W15.4S). Geomorphology. 2011;128:178-190. DOI:10.1016/j.geomorph.2011.01.004.

[33] Saillard M, Riotte J, Regard V, Violette A, Hérail G, Audin L, Riquelme R. Beach ridges ueth dating in Tongoy bay and tectonic implications for a peninsulaebay system, Chile. Journal of South American Earth Sciences. 2012;40:77-84. DOI: 10.1016/j.jsames.2012.09.001.

[34] Pfeiffer J. Evolución y génesis de calcretas pedogénicas en la paleobahía de Tongoy [Thesis]. Santiago: Universidad de Chile; 2011.

[35] Cabello M. Análisis geomorfológico de la sección occidental del Cordón Sarco: Identificación de terrazas marinas. Región de Coquimbo, Chile [Thesis]. Santiago: Universidad de Chile; 2015.

[36] Short AD. Waves-dominated beachs. In: Short, A. (ed). Handbook of beach and shoreface morphodynamics, Chichester: Wiley and Sons; 1999. P. 173-191.

[37] Soto M-V, Arriagada J, Cabello M. Geodinámica y tendencia evolutiva de Chile semiárido costero: la mega ensenada de Coquimbo. In: Borsdorf A, Marchant C, Rovira A, Sánchez R, editors. Handbook Chile cambiando. Revisitando la Geografía Regional de Wolfgang Weischet. Serie GeoLibros N° 36. 2020; p. 343-356. ISBN: 978-956-14-2718-1.

[38] Creixell T, Ortiz L, Arévalo C. Geología Del Área Carrizalillo – El Tofo. Servicio Nacional De Geología y Minería. Mapa geológico, N° 133 y 134, mapa escala 1:100.000, Santiago; 2012.

[39] Lagos G. Caracterización geomorfológica y dinámica costera de bahías del semiárido de Chile. Casos de estudio: Bahía Tongoy y Bahía Barnes, región de Coquimbo [Thesis]. Santiago: Universidad de Chile; 2013.

[40] Benavente N. Relaciones dinámicas asociadas al litoral-playa-duna anteriores del campo de dunas de Los Choros, región de Coquimbo [Thesis]. Santiago: Universidad de Chile; 2015.

[41] Rojas I. Caracterización dinámica de las dunas activas en la Ensenada de Los Choros, IV Región de Coquimbo [Thesis]. Santiago: Universidad de Chile; 2016.

[42] Soto MV, Sarricolea P, Sepúlveda SA, Rodolfi G, Cabello M, Maerker M. Assessment of hydro-geomorphological hazard potentials in the Chilean semiarid coastal range and its impacts on La Serena city, Coquimbo Region. Natural Hazards. 2017;88(1):431-452. DOI: 10.1007/s11069-017-2873-8.

Section 6

Coastal Hazards

Chapter 8

Storm Surge Risk Assessment for Non-Life Insurance

Rikito Hisamatsu

Abstract

This chapter introduces the efforts of the storm surge risk assessment for non-life insurance especially focusing on Japan. First, the importance of storm surge risk assessment in non-life insurance, the requirements for storm surge risk assessment in insurance, and an overview of the natural disaster model that evaluates them are described. Second, study on stochastic storm surge risk assessment, study on storm surge hazard modeling, study on vulnerability modeling which convert hazard intensity into damage are presented. Third, as an actual calculation example, the results of applying the procedure with low calculation load presented by past study to Tokyo Bay are shown. As a result, it is confirmed that the procedure can reduce the calculation load and maintain the calculation accuracy. Finally, how to select the existing storm surge risk assessment procedures when risk assessment is actually performed for the insurance purposes is considered.

Keywords: storm surge, risk assessment procedure, non-life insurance, stochastic typhoon model, Tokyo Bay

1. Introduction

This section introduces the natural disaster model and background of development in the insurance industry focusing on Japanese topics.

In 2016, the Japanese government revised the basic disaster management plan of Japan. According to the updated plan, Japanese government encourages the transfer of flood risk to non-life insurance because future flood risk will increase due to climate change [1]. Globally, economic losses and insurance losses due to natural disasters are on the rise, and the number of wind and flood damage that directly causes insurance losses is increasing significantly [2]. In Japan, on average, over 140 billion yen has been paid annually for insurance due to wind and flood damage since 1991 [3]. With the background of such government trends and increase of wind and flood risk, the risk transfer to non-life insurance has become more critical.

Non-life insurance companies measure the amount of natural disaster risk for their own risk management [4]. Therefore, it is important to properly evaluate the amount of risk. Storm surge damage with strong wind damage accompanying typhoons can be serious when the typhoon is highly intensified, and if damages of wind and storm surge occur at the same time, it may be a peak risk for non-life insurance companies [5]. The Ministry of Land, Infrastructure, Transport and Tourism's Port Bureau published "Guidelines for Storm surge Risk Reduction Policies on Offshore Areas of Ports" in 2018 [6]. Over 80% of the harbor areas in the three major bays of Japan (Tokyo Bay, Ise Bay, and Osaka Bay) are offshore areas of levee, and this guideline has promoted taking measures against storm surges on the areas. Even on the coasts of the three major bays, where the population and assets are particularly concentrated, small storm surges may cause floodwaters and damage many assets.

A major Japanese non-life insurance company has adopted 99.5% Value at Risk as the maximum risk amount that can be directly used for management decision-making [4]. In the natural disaster model that measures the risk amount with a low exceedance probability, hazards and economic loss are evaluated based on a probabilistic approach [7].

The history of the introduction of the natural disaster model in the insurance industry is following [8]. Around the 1960s, manual mapping was used for risk management until then, but information technology and Geographic Information System (GIS) have gradually advanced, and the technical base has been established. With regard to data and theory, scientific measurements of natural disasters have made rapid progress since the first half of the 20th century, and studies have been published that theorize the source and frequency of events by the 1970s. A computer-based model (a natural disaster model) for measuring a potential catastrophe was developed by fusing these technological bases, data and theory. Because of Hurricane Andrew in 1992, nine insurance companies could not pay perfectly for insurance claims, and necessity of the natural disaster model had been increased as a basis for decision making in many insurance companies and reinsurance companies. Along with the growing needs of the insurance industry, vendors of natural disaster models has grown and developed.

The natural disaster model used in the insurance industry consists of three modules: hazard module, vulnerability module, and financial module (**Figure 1**) [9]. In the natural disaster model, the hazard module first calculates wind speed, inundation depth, etc., then the vulnerability module converts the information into the damage ratio of the target object such as a house, and finally the financial module calculates the insurance loss considering insurance contract conditions. Natural disaster models continue to be important in insurance industry decision-making [10].

In underwriting, which examines insurance risks, the spatial distribution of inundation depth for each return period may be used [11]. However, the hazard map released by the Japanese government is a result based on the assumed scenario and is not sufficient to utilize it for the purpose of underwriting. Also, the expected annual loss calculated by the natural disaster model is used as a reference when insurance is priced [12].

From the above, the basic requirements to be satisfied by the natural disaster model from the perspective of non-life insurance are as follows.

1. To be able to predict low-frequency loss.

2. To be able to calculate the expected annual loss.

3. To be able to create a low-frequency inundation depth map that is used for underwriting.

This Chapter aims to discuss how to select the existing storm surge risk assessment procedures for non-life insurance. In order to discuss this point, review of the procedures including the latest methods with low calculation load and high calculation accuracy are introduced.

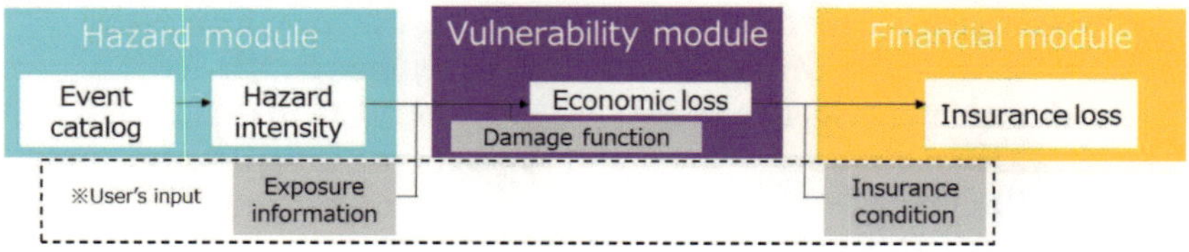

Figure 1.
Components of natural disaster model (based on InterRisk Research Institute & Consulting, Inc. [9] 2013).

2. Review of storm surge risk assessment methods

In this section, previous research and efforts related to storm surge risk assessment are described.

2.1 Study on stochastic storm surge evaluation

As a storm surge hazard study, Suzuki [13] conducted flood simulation over the whole of Japan, estimated the loss along the coast from the results, and developed loss function which is the relationship between water level at the representative point and the economic loss in the coastal area. The characteristics of this study are that the target area is wide and that large-scale storm surges due to climate change are also incorporated. However, since flood calculation is performed by a simple flood model, there is a issue in the accuracy of flood analysis, and no probabilistic discussion has been made in this study. Examples of publication on probabilistic storm surge damage estimation are following. Tsujita et al. [14] probabilistically calculated storm surge loss in the three major bays in Japan by using stochastic typhoon model. The stochastic typhoon model used in this study statistically processes past typhoon information and calculates the assumed typhoon that will occur in one year for 1000 patterns, so it does not consider future climate. The random variables are typhoon parameters (central pressure, location of typhoon, moving speed). This approach is common to the calculation of insurance purposes, which analyzes natural disaster risk in the coming year in many patterns and uses it for risk management and insurance premium setting. Exceedance probability used for risk management in non-life insurance companies is low. If the calculation period set long, the number of evaluation typhoons will increase and the calculation results will be stable and the uncertainty will decrease. The issue is that there is no discussion about the calculation period that was carried out. In addition, this study did not incorporate levees explicitly in the storm surge simulation, and no discussion on astronomical tide level setting. Similarly, Jiang et al. [15] probabilistically evaluated the current and future climate storm surge losses in the inner part of Ise Bay and showed the relationship between the annual exceedance probability and the storm surge loss. As for the present and future climates, a virtual typhoon for 25 ensembles of 200 years is used to evaluate the loss.

In the previous research mentioned above, it is common that firstly storm surge flood is calculated and secondly the loss is calculated from the inundation depth and the damage function which convert inundation depth into damage rate of the targeted assets. On the other hand, there is not enough discussion about uncertainty. In the process of calculating the storm surge loss, there are many parameters such as astronomical tide level, wind velocity/pressure distribution formula, maximum wind velocity radius, damage function, etc., and it is important to evaluate their effect on the estimated loss. Another issue is that there is no discussion about whether the calculation period is sufficient for the important return period.

A representative example of storm surge risk assessment efforts is the HUZUS-MH (Hazards U.S. Multi-Hazard) [16] developed by the Federal Emergency Management Agency (FEMA). This is software that estimates the damage caused by earthquakes, hurricanes, and floods in the United States, and displays damage of buildings and infrastructure due to past hazards. In addition, for the flood insurance program in the United States, FEMA has created a storm surge risk map with annual exceedance probability of 10, 2, 1, 0.2% in the United States, and has evaluated storm surge risk by various methods [17]. As a method, extreme value analysis, EST (Empirical Simulation Technique), and JPM (Joint Probability Method) have been studied. EST is a method of estimating the occurrence probability of the water level based on the observed tide level at a certain point by Probability distribution and constructing an artificial event set by the bootstrap method. JPM captures characteristics of hurricanes from observation information, constructs possible hurricanes from probability distributions of central pressure, moving speed and so on, and performs storm surge numerical calculation for all of them. While EST depends on limited observation data, JPM can comprehensively consider hurricanes etc. that may occur, and in recent years, JPM approach has been recognized as suitable for stochastic evaluation [17].

Similarly, in the natural disaster model of the insurance industry and probabilistic storm surge risk assessment in academic research, the method of calculating storm surges using assumed typhoons that capture past typhoon characteristics like JPM for a long term such as 1 year × 10,000 patterns has become common (eg AIRWORLDWIDE [18], Risk Management Solutions [19], Tsujita et al. [14]). That is, for example, when analyzing typhoons stochastically in Japan, it is assumed that a total of about 30,000 typhoons will land for 10,000 years because about 3 typhoons land in one year in average. However, it has been pointed out that the issue is that the computational cost is high because JPM calculates storm surges for all possible typhoons [17].

In order to reduce the calculation load of storm surge simulation, Jiang et al. [15] limited the typhoons that numerically calculate storm surge under the following three conditions.

- Typhoons that pass within 100 km of the bay
- Typhoons with a minimum central pressure of 950 hPa or less
- Typhoons with a typhoon speed of 20 km/h or more when landing

Through this process, typhoons that can flood are extracted. However, when considering variations in astronomical tide levels, it is not sufficient to evaluate storm surge risk because storm surge risk also depends on astronomical tide level, and it is necessary to use the total water level considering tides. In addition, the discussion on how much small and medium-scale storm surge damage can be extracted and uncertainty of annual expected loss of storm surge are insufficient in this paper.

In the US, FEMA has developed and is currently using JPM-OS (Joint Probability Method – Optimal Sampling), which is a method that reduces the calculation load of JPM (Johnson et al. [20]). Here, JPM-OS is briefly described. First, JPM-OS performs storm surge inundation analysis only on representative events selected from the numerous typhoon events. Then, the results are interpolated to estimate the inundation depth of the event for which no flood analysis has been performed. Although it is possible to reduce the calculation load by one digit compared to JPM [17], the problem is that uncertainty arises during interpolation. In order to reduce the uncertainty, research is underway on JPM-OS interpolation methods (eg Yang et al. [21]). Yang et al. compared the calculation results of the inundation depth for different JPM-OS interpolation method for each return period in Florida, USA. The RMSE of each method was 0.16 to 0.82 m when the inundation depth at each point was compared with the calculation result by JPM for each interpolation method for the return period of 50, 100, and 500 years. It was shown that an error occurs between the interpolated result and the numerical calculation result regardless of whether the return period is long or short.

As another approach, Hisamatsu et al. [5] selected a typhoon that can be inundated by a simple formula for typhoons of the stochastic typhoon model, and calculated the storm surge inundation only for the selected typhoon using a numerical model. Procedure by Hisamatsu et al. is described in **Figure 2**. This method aims at both reduction of calculation load and preservation of calculation accuracy by extracting floodable typhoons. Additionally, fluctuation of astronomical tide can be considered. The brief results of applying this procedure to the Tokyo Bay, Japan are shown in Section 3.

2.2 Study on hazard modeling

There are various storm surge models all over the world, but here some of them are introduced.

As a method for numerical analysis of storm surge hazards, a lot of studies have shown that considering wave set-up improves reproducibility. Kim et al. [22] developed a SuWAT (Surge-Wave-Tide coupled model), which is a model that considers wave set-up. SuWAT is a two-way coupled model that considers the interaction between tidal, storm surge and wave. The SuWAT model, composed of depth-integrated nonlinear shallow water equations and a simulated-waves near-shore (SWAN) model, can simultaneously run an arbitrary number of nested domains by using the message passing interface (MPI). Mase et al. [23] simulated typhoon Vera 1959 using SuWAT, and showed that the reproducibility of the storm surge is greatly improved if wave set-up is incorporated explicitly. Since SuWAT has been used by a lot of research especially in Japan and other Asian regions (eg Hisamatsu et al. [5]), this model has been adopted to natural disaster model of the insurance industry as numerical simulation model for storm surge [24].

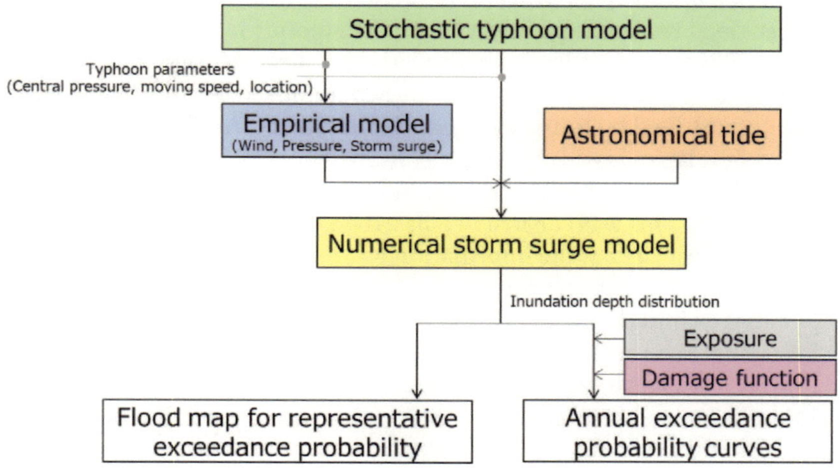

Figure 2.
Procedure of stochastic storm surge proposed by Hisamatsu et al. (based on Hisamatsu et al. [5] 2020).

In flood calculation, models that improve the calculation speed have been used. For example, Ramirez et al. [25] used LISFLOOD-FP [26] for flood calculation, which is dynamic and has a small calculation load. Inundation calculation load of storm surge was further reduced by using the flood model with the results of the storm surge model as boundary conditions (eg. Tsujita et al. [14]).

2.3 Study on vulnerability modeling

Some existing research on the damage function, which estimates the damage of assets from the inundation depth, are presented.

Regarding the relationship between tsunami inundation depth and its damage, as a representative of the tsunami damage caused by the 2011 Great East Japan Earthquake, the fragility curves expressing the probability of occurrence for each degree of damage have been published (eg Suppasri et al. [27], Aránguiz et al. [28]). These fragility curves are functions that calculate the occurrence probability P_i of each damage level i (minor, moderate, major, complete) using the inundation depth η as an explanatory variable as shown in the following equation (Eq. (1)).

$$P_i = f_i(\eta) \tag{1}$$

However, since this fragility curve does not show the damage ratio of assets, it is not possible to directly calculate the loss from the hazard intensity of the tsunami. Dias et al. [29] presented a method of converting the fragility curves into damage function. In other words, the tsunami fragility curves that have been accumulated so far can be converted into a tsunami damage function that can directly calculate the asset damage ratio R using the inundation depth as an explanatory variable, as shown in the following equation (Eq. (2)).

$$R = f(\eta) \tag{2}$$

The storm surge damage function of HAZUS in the United States is frequently used to estimate the storm surge damage amount, which converts the storm surge inundation depth into asset damage rate (eg, Johnson et al. [20], Lin et al. [30]). The damage function installed in HAZUS was developed by US Army Corps of Engineers through post-flood research and interviews with experts [31, 32]. In addition, Kar and Hodgson [33] theoretically constructed a storm surge damage function.

In Japan, damage functions in Manual of Economic Survey for Water Management published by Ministry of Land, Infrastructure, Transport and Tourism (MLIT), Japan is widely used (eg Hisamatsu et al. [5], Tsujita et al. [14], Jiang et al. [15]). The reason is that there is no other storm surge damage function for Japanese assets. This was constructed based on the survey conducted in 1993 to 1996, and the issue is that the information of the material and equipment of the house, etc. surveyed deviates from present. Therefore, some study described relationships between inundation depth and damage ratio based on survey and simulation as following. Suzuki et al. developed flood damage function by hearing-based survey [34]. And Hisamatsu et al. developed flood damage function using the result of flood simulation and insurance data [35]. Unfortunately, small number of studies on damage functions is conducted. However, it is possible to accumulate storm surge damage functions based on the approach described above for actual events.

The shape of the damage function is roughly divided into two types, a step function and a continuous function. Kar and Hodgson [33] and the MLIT are the former, while US Army Corps of Engineers [32], Suzuki et al. [34] and Hisamatsu et al. [35] are the latter. In other cases, such as Tsujita et al. [14], the step function of the MLIT is regressed and converted into a continuous function and is used for damage estimation. In the case of the step function, since the damage function is constructed by setting the damage of modeled building according to the inundation depth, the function that the damage increases when the water level reaches the floor or ceiling of the modeled building. However, the structures of buildings that are actually damaged vary, damage ratio at the same inundation depth differs depending on the building. Therefore, the damage functions of US Army Corps of Engineers [32] and Suzuki et al. [34], which were constructed based on the disaster survey, is a continuous damage function because it covers buildings with various structures.

3. Case study in the Tokyo Bay, Japan

The storm surge risk assessment procedure described in **Figure 2** was applied to the Tokyo Bay where assets are concentrated in Japan in order to whether the calculation accuracy can be maintained by reducing the calculation load is considered.

As a stochastic typhoon model, global stochastic typhoon model (GSTM) developed by Nakajo et al. [36] is used for evaluation. Typhoons created by the GSTM is extracted around Tokyo Bay (**Figure 3**), and top 1000 typhoons are used as input data of numerical model SuWAT following proposed procedure. Prior to apply SuWAT, reproductivity of the model is validated by calculating time series storm surge levels of Typhoon Irma along the Tokyo Bay coast. The astronomical tide level

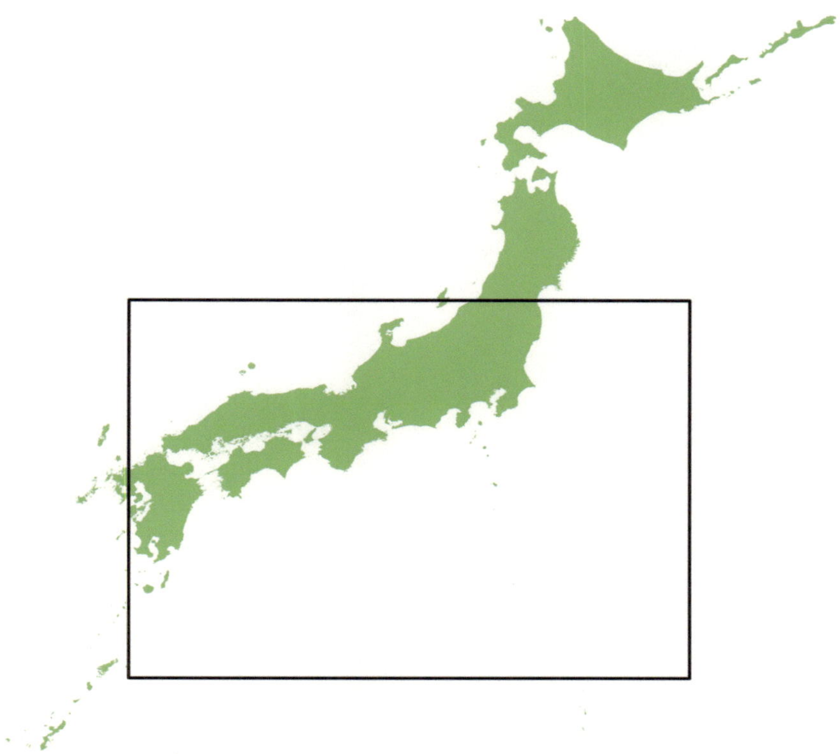

Figure 3.
Area where typhoon is extracted (Hisamatsu et al. [37] 2020).

was calculated by using the harmonic constants to estimate the time series tide level at the Tokyo tidal station for 100 years from January 2000 to December 2099. The astronomical tide level for each typhoon was set by randomly extracting from this histogram of calculated astronomical tide levels. From above calculation, storm surge inundation depth distributions due to 1000 typhoons are obtained. Example of the results is shown in **Figure 4**.

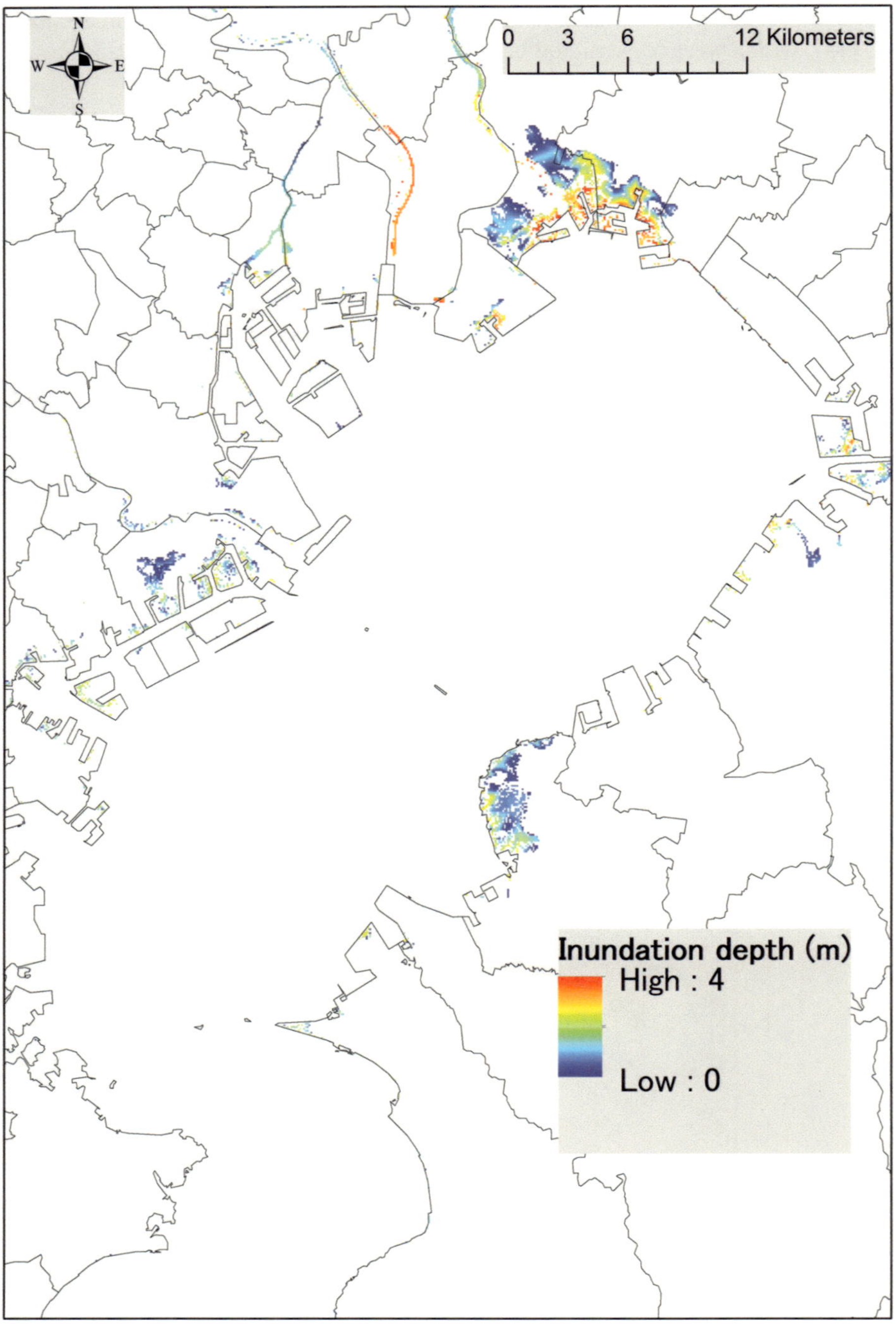

Figure 4.
Example of simulation result.

In addition, economic loss is estimated by using the inundation depth calculated and damage functions by MLIT. Targeted assets are houses and business establishments and loss calculation consider number of floors. By using calculated loss amount, exceedance probability curve is created as described in **Figure 5**.

From the above calculations, it was confirmed that the insurance requirements shown in Section 1 are satisfied by applying suggested procedure. However, it is necessary to consider whether or not calculation accuracy can be maintained by reducing the calculation load following the procedure. Here, consideration conducted by Hisamatsu et al. [37] is introduced.

The storm surge loss introduced in this section was estimated by extracting typhoons from the stochastic typhoon model based on the top 1000 water levels by the storm surge empirical formula. The maximum water level by the storm surge empirical formula used in the extraction process is different from the numerical model result. Therefore, the ranking of the maximum water level in the Tokyo Bay differs between the numerical model and the empirical formula. It is important to check whether the number of events extracted by empirical formula was sufficient for insurance purposes, in order to confirm the usefulness of the proposed procedure. **Figure 6** shows the estimated loss by rank for each number of extracted typhoons. The horizontal axis shows the number of typhoons extracted from the top of the total water level based on the empirical formula, in other words, number of typhoons used for the analysis, and confirms how many typhoons the loss amount of the target order will converge. In the analysis based on 1000 typhoons, the losses in the top 50 and above were almost converged. It was suggested that 1000 typhoons based on the proposed procedure are generally sufficient to obtain low-frequency damage amounts for the purpose of insurance. In addition, it was confirmed that the infrequent water levels in Chiba, Yokohama, and Yokosuka would converge with 1000 typhoons extracted in the same way.

According to discuss the annual expected loss amount, it was confirmed that the annual expected loss amount not being fully converged. It was found that the reason was the creation of the asset amount distribution. Since the amount of assets is created from statistical information, the resolution is coarser than that of storm surge numerical analysis. Because the statistical information is distributed according to the resolution of the numerical analysis, it is distributed to the place where the asset originally does not exist and the loss amount is calculated.

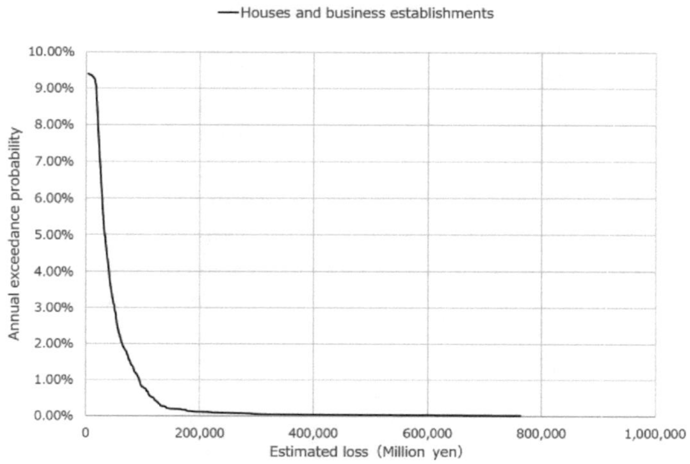

Figure 5.
Annual exceedance probability curve.

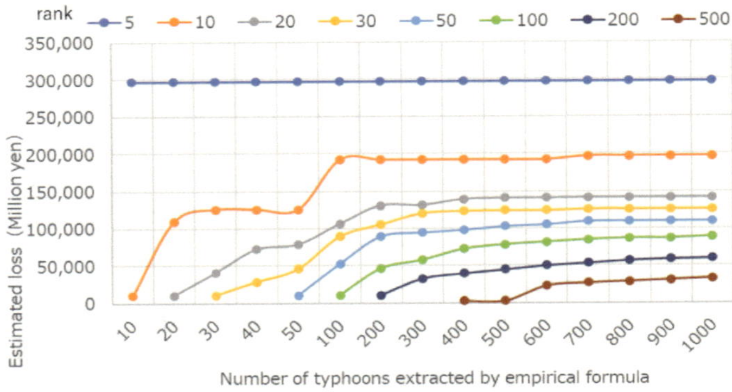

Figure 6.
Estimated loss by rank for each number of extracted typhoons (based on Hisamatsu et al. [37] 2020).

On the other hand, the places where the number of floods was extremely high in the numerical calculation of storm surges are waterside areas where no assets actually exist. As a result of estimating the loss amount ignoring the loss at these areas, it was confirmed that the expected annual loss amount has converged. Therefore, if the asset amount distribution can be corrected more realistic this problem will be solved. It also suggests that asset allocation is very important for risk assessment.

4. Conclusion

In this chapter, the efforts of storm surge risk assessment in the non-life insurance industry are introduced based on the author's experience. In the insurance industry, probabilistic storm surge risk assessment is required for risk management and underwriting, and a lot of analysis is required. Therefore, how to reduce the calculation load without degrading the calculation accuracy is being discussed. In this chapter, author introduced a previous study on reducing computational load. In particular, an example in which the procedure of numerically calculating only typhoons that can cause floods was applied to Tokyo Bay is introduced.

This procedure is useful in two ways. The first point is to significantly reduce the calculation load. The procedure aims at both the reduction of the calculation load and the reduction of the calculation error compared to other methods. By applying the procedure, it was found that for Tokyo Bay, the number of typhoons to be calculated can be reduced from about 90,000 to 1000 and the calculation accuracy can be maintained. By utilizing this procedure, the cost of storm surge risk assessment can be reduced, and the reduction of insurance rate may be possible, so that taking out insurance and transferring risks to ensure a safer life for more people are expected. In addition, even if the stochastic typhoon model is updated, the risk can be quickly evaluated based on the latest knowledge considering climate change and reflected in the risk management of the insurance company. The second point is that the loss amount can be evaluated more appropriately by considering the variation of the astronomical tide level. The uncertainty associated with risk assessment for insurance purposes can be recognized and reduced by varying the astronomical tide level.

In the insurance industry, the procedure applied to the Tokyo Bay in this chapter and the method like JPM-OS should be used separately. First, the characteristics of the evaluation target site should be considered. Since the procedure applied to the Tokyo Bay analyzes storm surges for all typhoons that may flood, it has a great effect on reducing computational load in areas where flooding is unlikely to occur, such as the Tokyo Bay. However, when targeting areas with high flooding frequency, such as Southeast Asia, the number of typhoons that can be flooded will be huge, and the effect of reducing computational load cannot be expected. Therefore, it is necessary to judge the risk assessment by JPM-OS allowing the calculation accuracy in such areas with high flood frequency. Next, requirements to be evaluated should be considered. It is necessary to analyze a lot of typhoons, especially when obtaining the expected annual loss. The distribution of inundation depths for each return period used for underwriting and the loss for the representative return period required for risk management target at low-frequency risks, so it is not necessary to analyze all typhoons of stochastic typhoon model. In this case, if the method applied to the Tokyo Bay in this chapter is used to calculate until the hazard and loss amount in the representative return period converge, it is possible to reduce the calculation load and evaluate with high accuracy.

Acknowledgements

The author sincere thanks Professor Shigeru Tabeta and Assistant professor Katsunori Mizuno of The University of Tokyo for their considerable guidance. Special thanks are due the members of MS&AD InterRisk Research and Consulting, Inc. for their comments and supports.

Conflict of interest

The authors declare no conflict of interest.

Author details

Rikito Hisamatsu
MS&AD InterRisk Research and Consulting, Inc., Tokyo, Japan

*Address all correspondence to: rikito.hisamatsu@ms-ad-hd.com

References

[1] Cabinet office, Government of Japan. Basic disaster management plan [Internet]; 2017. Available from: http://www.bousai.go.jp/taisaku/keikaku/pdf/kihon_basic_plan170411.pdf [Accessed: 2020-9-23]

[2] MunichRe. TOPICS GEO 2017; 2018. P.53

[3] The General Insurance Association of Japan. FACTBOOK 2019; 2019. P.77

[4] Workshop of Enterprise Risk Management. Theory and Practice of Enterprise Risk Management for Insurance; 2015. Kinzai Institute of Financial Affairs, Japan.

[5] Hisamatsu R, Tabeta S, Kim S, Mizuno K. Storm surge risk assessment for the insurance system: A case study in Tokyo Bay, Japan. Ocean and Coastal Management. 189. 2020; DOI: https://doi.org/10.1016/j.ocecoaman.2020.105147

[6] The Ministry of Land, Infrastructure, Transport and Tourism's Port Bureau. Guidelines for Storm surge Risk Reduction Policies on Offshore Areas of Ports; 2018. Available from: https://www.mlit.go.jp/common/001228806.pdf [Accessed: 2020-9-23]

[7] Aon Benfield Japan. Survey on the structure of an external procurement model for natural disaster risk; 2012. Available from: http://www.fsa.go.jp/common/about/research/20120706/01.pdf [Accessed: 2020-9-23]

[8] General Insurance Rating Organization of Japan. Catastrophe Modelling -New approach of risk management-. Earthquake insurance report; 2006.

[9] InterRisk Research Institute & Consulting, Inc. ; Development of Probabilistic Tsunami risk model; 2013. Available from: https://www.cybernet.co.jp/avs/documents/pdf/seminar_event/conf/19/3-3.pdf [Accessed: 2020-9-23]

[10] Karaca E, Aslani H. Review of two Japan Typhoon catastrophe models for commercial and industrial properties. Nat Hazards 83, 19-40. 2016; https://doi.org/10.1007/s11069-016-2340-y

[11] AIRWORLDWIDE. AIR probabilistic Flood Hazard Map in Touchstone; 2018. Available from: https://www.air-worldwide.com/siteassets/Publications/Brochures/documents/AIR-Probabilistic-Flood-Hazard-Maps-in-Touchstone [Accessed: 2020-9-23]

[12] General Insurance Rating Organization of Japan. Research on risks of natural disaster ; Available from: https://www.giroj.or.jp/databank/natural_disaster.html [Accessed: 2020-9-23]

[13] Suzuki T. JAPAN'S INNUNDATION RISK MAP OF STORM SURGE UNDER GLOBAL WARMING AND INNUNDATION DAMAGE FUNCTION OF COASTAL AREA. Journal of Japan Society of Civil Engineers, Ser. B3 (Ocean Engineering), Vol. 68, Issue 2, I_870-I_875. 2013; DOI: https://doi.org/10.2208/jscejoe.68.I_870

[14] Tsujita D, Yasuda T, Shinohara M, Mori N, Mase H. Evaluation of aggregate loss by storm surge considering simultaneous damage occurrence in multiple areas. J. Jpn. Soc. Civ. Eng. B2 (Coast Eng.) 72 (2), I_1639–I_1644. 2016; DOI: https://doi.org/10.2208/kaigan.72.I_1639

[15] Jiang X, Mori N, Tatano H, Yang L. Simulation-Based Exceedance Probability Curves to Assess the Economic Impact of Storm Surge

Inundations due to Climate Change: A Case Study in Ise Bay, Japan. Sustainability. 11. 1090. 2019; DOI: 10.3390/su11041090

[16] Federal Emergency Management Agency. Huzus; Available from: https://www.fema.gov/hazus [Accessed: 2020-9-23]

[17] Federal Emergency Management Agency. Operating Guidance No.8-12 For use by FEMA staff and Flood Hazard Mapping Partners; Available from: https://www.fema.gov/media-library-data/2262de940ad81df454aace5228a1a35d/24.+operating_guidance_8_12.pdf [Accessed: 2020-9-23]

[18] AIRWORLDWIDE. AIR Typhoon Models for Southeast Asia; Available from: https://www.air-worldwide.com/SiteAssets/Publications/Brochures/documents/AIR-Typhoon-Models-for-Southeast-Asia [Accessed: 2020-9-23]

[19] Risk Management Solutions. RMS Philippines Typhoon and Inland Flood Model; Available from: https://forms2.rms.com/rs/729-DJX-565/images/rms-philippines-typhoon-model-datasheet.pdf [Accessed: 2020-9-23]

[20] Johnson D R, Fischbach J R, Ortiz D S. Estimating Surge-Based Flood Risk with the Coastal Louisiana Risk Assessment Model. Journal of Coastal Research. 67. 109-126. 2013; DOI: https://doi.org/10.2112/SI_67_8

[21] Yang K, Paramygin V, Sheng Y P. An objective and efficient method for estimating probabilistic coastal inundation hazards. Natural Hazards. 99. 1105-1130. 2019; DOI: https://doi.org/10.1007/s11069-019-03807-w

[22] Kim S, Yasuda T, Mase H. Numerical Analysis of Effects of Tidal Variations on Storm Surges and Waves. Applied Ocean Research. Vol. 30. No. 4. 311-322. 2008; DOI: https://doi.org/10.1016/j.apor.2009.02.003

[23] Mase H, Muto R, Mori N, Kim S, Yasuda T, Hayashi Y. Storm Surge Simulation due to Isewan Typhoon using Detailed Meteorological Re-analysis Data. Journal of Japan Society of Civil Engineers, Ser. B2 (Coastal Engineering). Vol. 67. 2. I_401-I_405.2011; DOI: https://doi.org/10.2208/kaigan.67.I_401

[24] AIRWORLDWIDE. A Preview of the AIR Typhoon Models for Southeast Asia; Available from: https://www.air-worldwide.com/SiteAssets/Publications/Presentations/attachments/A-Preview-of-the-AIR-Typhoon-Models-for-Southeast-Asia [Accessed: 2020-9-23]

[25] Ramirez J, Lichter M, Coulthard T, Skinner C. Hyper-resolution mapping of regional storm surge and tide flooding: comparison of static and dynamic models. Natural Hazards. 82. 571-590.2016; DOI: https://doi.org/10.1007/s11069-016-2198-z

[26] Schumann G -P, Neal J, Voisin N, Andreadis K, Pappenberger F, Phanthuwangpakdee N, Hall A, Bates P. A first large-scale flood inundation forecasting model. Water Resources Research. 49. 6248-6257. 2013; DOI: https://doi.org/10.1002/wrcr.20521

[27] Suppasri A, Mas E, Koshimura S, Imai K, Harada K, Imamura F. Developing Tsunami Fragility Curves from the Surveyed Data of the 2011 Great East Japan Tsunami in Sendai and Ishinomaki Plains. Coastal Engineering Journal. 54:1. 1250008-1-1250008-16. 2012; DOI: 10.1142/S0578563412500088

[28] Aránguiz R, Urra L, Okuwaki R, Yagi Y. Development and application of a tsunami fragility curve of the 2015 tsunami in Coquimbo, Chile. Natural Hazards and Earth System Sciences.

18. 2143-2160. 2018; DOI: https://doi.org/10.5194/nhess-18-2143-2018

[29] Dias W, Edirisooriya U. Derivation of tsunami damage curves from fragility functions. Natural Hazards. 96. 1153-1166. 2019; DOI: https://doi.org/10.1007/s11069-019-03601-8

[30] Lin N, Shullman E. Dealing with hurricane surge flooding in a changing environment: part I. Risk assessment considering storm climatology change, sea level rise, and coastal development. Stochastic Environmental Research and Risk Assessment. 31. 2379-2400. 2017; DOI: https://doi.org/10.1007/s00477-016-1377-5

[31] Federal Emergency Management Agency. Hazus Flood Model User Guidance; Available from: https://www.fema.gov/media-library-data/1564766454464-d77d2c219be0f54315aa79ac5dbc3547/Hazus_4-2_Flood_User_Manual_August_2019.pdf [Accessed: 2020-9-23]

[32] US Army Corps of Engineers. North Atlantic Coast Comprehensive Study: Resilient Adaptation to Increasing Risk; Available from: https://api.army.mil/e2/c/downloads/379567.pdf [Accessed: 2020-9-23]

[33] Kar B, Hodgson M E. Observational Scale and Modeled Potential Residential Loss from a Storm Surge. GIScience & Remote Sensing. 49. 202-227. 2012; DOI: https://doi.org/10.2747/1548-1603.49.2.202

[34] Suzuki S, Koshimura S, Harada K, Okamoto M, Fukutome K, Suga M, Kawata Y. DEVELOPMENT OF FRAGILITY FUNCTION FOR WOODEN HOUSE OBTAINED FROM THE FIELD INVESTIGATION OF FLOOD DISASTER DUE TO JULY 13, 2004 NIIGATA HEAVY RAINFALL. Journal of Japan Society of Civil Engineers, Ser. B1 (Hydraulic Engineering). 49. 439-444. 2005; DOI: https://doi.org/10.2208/prohe.49.439

[35] Hisamatsu R, Kawabe K, Mizuno Y, Shinozuka Y. Horie K. DEVELOPMENT OF FLOOD DAMAGE FUNCTIONS BASED ON INSURANCE LOSS DUE TO 2015 KANTO–TOHOKU HEAVY RAINFALL. Journal of JSCE. Vol.7. 22-29. 2019; DOI: https://doi.org/10.2208/journalofjsce.7.1_22

[36] Nakajo S, Mori N, Yasuda T, Mase H. Global Stochastic Tropical Cyclone Model Based on Principal Component Analysis and Cluster Analysis. J. Appl. Meteor. Climatol.. 53. 1547-1577.2014; DOI: https://doi.org/10.1175/JAMC-D-13-08.1

[37] Hisamatsu R, Tabeta S, Kim S, Mizuno K. CONSIDERATION ON UTILITY OF STOCHASTIC STORM SURGE RISK ASSESSMENT PROCEDURE FOR REDUCING CALCULATION LOAD. Journal of Japan Society of Civil Engineers, Ser. B1 (Hydraulic Engineering), vol. 76, 2020

Chapter 9

Development of an Ocean Hazards Classification Scheme (OHCS) for Projecting Future Scenario Vulnerability Ranking on Coastal Built Infrastructure

Oceana Francis, Linqiang Yang, Harrison Togia and Gleb Panteleev

Abstract

From many sources, we develop an ocean hazard classification scheme (OHCS) based on the collection of historical and projected ocean hazards data at 302 locations along Hawaii's state coastal highways. The OHCS identifies ocean hazards impacting coastal built infrastructure, i.e. roadways. In the OHCS, we first rank the vulnerability of: sea level rise; waves; shoreline change; tsunami; and storm surge. Next, using our developed OHCS, provide the vulnerability ranking for all five variables combined. We find the highest OHCS to be on Molokai, the island that has the highest OHCS numbers for most of the island. For the majority of state highway locations in Hawaii, we find the highest vulnerability is from storm surge, with tsunami threat being the second largest contributor. Sea level rise should also be considered a contributor since higher sea levels contribute to more extreme storm surge and tsunami inundation. Although the OHCS is applied towards roads in our study, our method can be applied towards any coastal island-based built infrastructure vulnerability scheme. This is an important tool in planning for future construction projects or identifying which hazards to focus on in more detailed assessments, such a probabilistic risk assessment in a more localized location.

Keywords: ocean hazards, vulnerability ranking, Hawaii statewide highways, sea level change rate, wave height, shoreline change rate, tsunami inundation, storm surge inundation

1. Introduction

Throughout the northern and southern Pacific Oceans, lay many remote islands. These islands are prone to extreme waves, tectonic activity, and climate change which results in storm surges, shoreline change, tsunamis, and sea level rise. The remoteness of these islands, which allows these regions to capture fully-developed seas, and their lack of a continental shelf, puts them at particular risk to ocean hazards. The Hawaiian Islands are among the most remote islands in the world. Seven natural phenomena have been identified as posing significant threat to coastal areas of the Hawaiian Islands which include: coastal erosion, sea level rise, major storms, volcanic and seismic activity, tsunami inundation, coastal stream flooding, and extreme seasonal high wave events [1]. Coastal slope, distance to shoreline and geologic setting are also important factors when considering coastal infrastructure exposure and vulnerability.

We consider ocean hazards on coastal infrastructure, in this case, road infrastructure. Our previous study [2] used ocean hazard values which include: historical sea level rise, historical significant wave height, tides, and historical shoreline change (without sea level rise). A methodology was developed to quantify historical ocean hazards at critical road locations that have particularly large CRESI (Coastal Road Erosion Susceptibility Index) values and where the Department of Transportation is concerned about road collapse. Note, that although tides is an important ocean variable that should correctly be considered around much of the world, we have omitted it now, in this study, since the Hawaiian Islands have a low mean tidal range of about 2 feet [3].

Here, we propose that using projected ocean hazards may give a more accurate representation when planning for future climate change on infrastructure. The ocean hazards we use include historical and projected sea level rise; projected shoreline change with sea level rise; projected storm surge; historical and projected tsunamis, and historical extreme seasonal high wave events. We develop a quantitative Ocean Hazards Classification Scheme (OHCS) based on the Ocean Hazards Database (OHD) [4] of 302 mileposts across coastal state routes in Hawaii (**Figure 1**). These

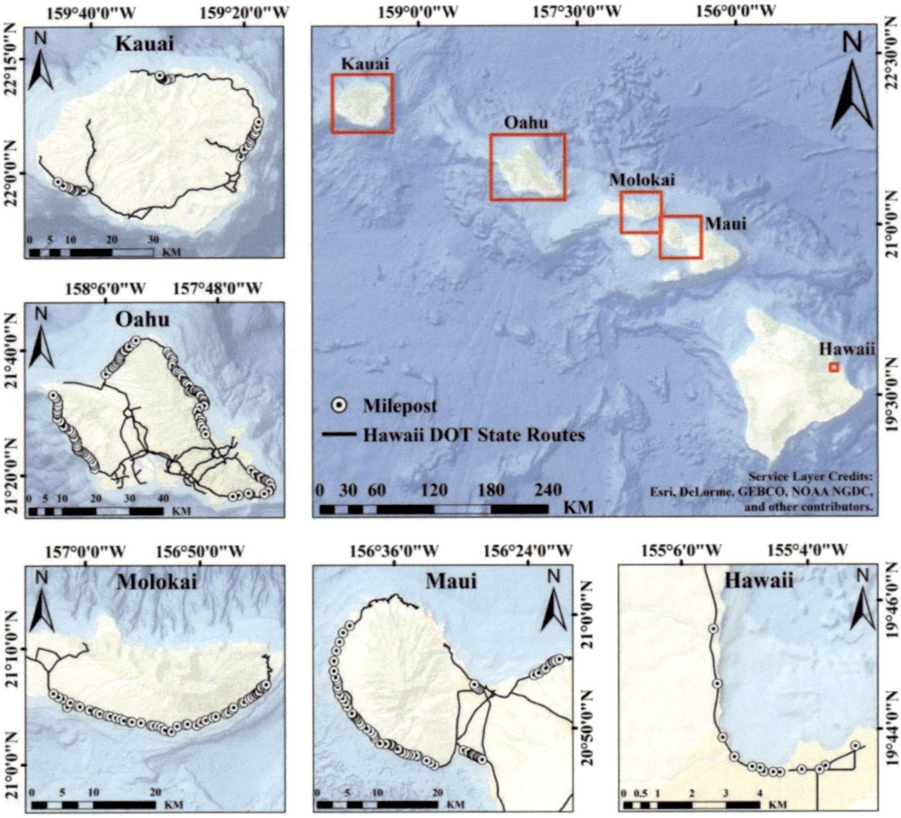

Figure 1.
Study location area, State of Hawaii, USA. The red squares show the location on each island. The white circles (with inner black dots) indicate the milepost (MP) locations where each measurement was taken.

mileposts are identified as vulnerable in [2] due to: road distance to the shoreline, road elevation, and historical road degradation due to coastal processes.

Although probability risks assessments (PRAs) are used widely to give predictions of storminess or shoreline change in a region, it is a time consuming method requiring historical data for twenty years or more, in order to create accurate projections. Also, it is often limited to a localized area due to long computational times.

The aim of this study is to obtain the projected ocean hazards vulnerability rankings using projected rates and projected inundation and the CVI method [5]. The data comes from various governmental and academic sources, which are used and put into the OHCS equation to develop one number, ranging from 0 to 100. With these rankings, we are not only able to produce an overall vulnerability ranking for the five hazards, but we are also able to identify which of the five hazards most affects the coastal road section in a region.

In the next section, we discuss our methodology and the development of the five ocean hazard variables (sea level rise, waves, shoreline change, tsunamis, storm surge) that we use. In Section 3, we give the results by evaluating which of the five ocean hazards most affect vulnerable highway sections in the State of Hawaii and show the overall vulnerability rankings. In the subsequent sections, Sections 4, 5, 6, 7, and 8, we give the conclusions, acknowledgements, references, figures and tables, respectively.

2. Methods and data

2.1 Methodology

The use of historical and projected values is important towards the development of an Ocean Hazards Classification Scheme (OHCS) for projecting future scenario vulnerability ranking on coastal built infrastructure. Our variables we consider: (1) sea level rise rate, (2) wave height, (3) shoreline change rate, (4) tsunami inundation, and (5) storm surge inundation, are described here.

Variable (1), sea level rise, is the sea level rise rate (1905–2050, extreme scenario) (in/yr). Local sea level rise is the result of both global sea level rise and local factors. Global sea level rise is due to warmer ocean temperatures and melting land ice, both caused by climate change. Local factors include land motions and tides, currents, and winds. Local sea levels can rise faster than the average global rate.

Variable (2), maximum annually recurring waves, is the significant wave height (2010–2018) (ft). This includes all forecasted wind-waves from 2010 to 2018, which was modeled in the wind-driven Simulating WAves Nearshore (SWAN) wave model.

Variable (3), shoreline change, is the mean projected shoreline change rate (2008–2100) (ft/yr); and CRESI – armoring ranking (1–5) [6]. Variable 3, shoreline change, determines the seaward encroachment of the beach towards the road and how protected the road is, whether there is existing armoring or not. Shoreline change is seasonal, where erosion and accretion are present during different times of the year. The most significant shoreline change is influenced by wave action, particularly storm surge events, which occur almost annually, transporting much of the coastline away during one event.

Variable (4), tsunamis, considers the historical and hypothetical inundation (ft). Variable 4, tsunamis, are seismic ocean waves causing coastal inundation caused by earthquakes, underwater landslides, volcanic eruptions, or meteorites.

Variable (5), storm surge, is Category 1, 2, 3, and 4 storm inundation (ft). Variable 5, storm surge, is a rapid rise in sea level causing coastal inundation due to low pressure, high winds, and high waves associated with hurricanes.

Using the ranking, from 1 to 5, for each of the five variables, we input these variables into one equation (Eq. (1)), which we call the Ocean Hazard Classification Scheme (OHCS), to obtain a value between 1 to 100, where the higher the values, the more vulnerable the region.

$$OHCS = \sqrt{\frac{(Variable\,1 * Variable\,2 * Variable\,3 * Variable\,4 * Variable\,5)^{1.345}}{5}} \quad (1)$$

where *Variable* 1 is 2050 sea level rise rate ranking (extreme scenario) (1 to 5), *Variable* 2 is significant wave height ranking (1 to 5), *Variable* 3 is mean shoreline change rate ranking (1 to 5), *Variable* 4 is tsunami inundation ranking (1 to 5), and *Variable* 5 is storm surge inundation ranking (1 to 5).

Eq. (1) is taken as the square root of the geometric mean of the ranking variables, with the addition of a power scalar to adjust the range of theoretical OHCS rankings to maximize at a value of 100. Therefore as the number of variables change, so does the scalar power. When considering five input variables, each with a maximum ranked value of 5, a power scalar value of 1.345 results in a potential maximum OHCS value of 100. Our method is similar to that used in Chapter 1 of [2] for calculating Coastal Road Erosion Susceptibility Index (CRESI) values, in [7, 8] who was the first to use the coastal vulnerability index (CVI) for the entire Hawaiian Islands to assess coastal vulnerability, and that described by [5] for finding the coastal vulnerability index (CVI) rankings.

2.2 Sea level rise

Historical rates of sea level rise are estimated from observed data, and future sea level rise rates are estimated from projected data. For both historical and future scenarios, it is essential to take the spatial variation into consideration when determining the rate of sea level rise. For this reason, we divide each island into a certain number of segments and derive the historical and future sea level rise rates for each segment, respectively. Currently, there are two types of data used to estimate the historical sea level rise rate: tide gauge and satellite altimetry data. Tide gauges are usually placed on piers and measure the sea level relative to a nearby geodetic benchmark, known as relative sea level (RSL). Satellite altimetry measures the sea level relative to a reference ellipsoid, known as absolute sea level (ASL). Here, we study how the sea level rise affects the coastal infrastructure (i.e. roads) in the Hawaiian Islands. Therefore, we focus on the trend estimates of RSL. There are six tide gauge stations in operation in the Hawaiian Islands: NAWI is located in Nawiliwili Bay, Kauai Island with data spanning 1955–2016; MOKU is located in Mokuoloe Island, Oahu Island with data spanning 1957–2016; HONO is located in Honolulu, Oahu Island with data spanning 1905–2016; KAHA is located in Kahului Harbor, Maui Island with data spanning 1947–2016; KAWA is located in Kawaihae, Hawaii Island with data spanning 1988–2016; and HIHA is located in Hilo, Hawaii Island with data spanning 1927–2016. The RSL data of the six available stations in the Hawaiian Islands are downloaded from the Permanent Service for Mean Sea Level (PSMSL) [9, 10]. We make use of all available RSL data from the six tide gauge stations to estimate the RSL trends, respectively. Before estimating the RSL trends, the following process is applied. First, the seasonal signal is removed from the RSL time series using the Seasonal Trend Decomposition using Loess (STL) procedure [11]. Second, we remove the common-mode-oceanographic signals from each RSL time series. The common-mode-oceanographic signals can be derived by averaging the monthly detrended and de-seasoned RSL time series of the all six available tide-gauge stations in the Hawaiian Islands. Finally, the linear trends of the RSL are estimated. However, tide gauge stations are sparsely distributed and not all the segments are covered. For those segments not covered by the tide gauge stations, an indirect way is applied to derive the relative sea level rise trend (RSLT). The RSL variation is comprised of two components: ASL variation and vertical land motion (VLM). Eq. (2) indicates the relationship of the three components:

$$ASLT = RSLT + VLMR \tag{2}$$

where $ASLT$ represents ASL trend, $RSLT$ represents RSL trend, and $VLMR$ represents VLM rate. Therefore, the RSLT of the segments without tide gauge stations can be estimated by combining the ASLT and VLMR. In this paper, we use the reprocessed and merged-gridded sea-level-anomaly heights for global areas processed by Ssalto/Duacs [12] to derive the ASLT. The satellite altimetry data spans 1993–2017 and has a resolution of 0.25 arc degrees. If there is more than one satellite altimetry grid point near the study segment, the time series are averaged to derive the ASLT. Before estimating the ASLT, the Dynamic Atmospheric Correction (DAC) is downloaded and added back to the satellite altimetry data to keep in accordance with the tide gauge data which do not use the barometric pressure correction. The DAC data are produced by Collecte Localisation Satellites (CLS) using the Mog2D model from Legos and distributed by Aviso+, with support from CNES (https://www.aviso.altimetry.fr/). The satellite altimetry data is accessible at the Copernicus Marine Environment Monitoring Service (CMEMS) (http://marine.copernicus.eu/). The data of Global Navigation Satellite System (GNSS) which has proven to be a robust tool to monitor VLM [13–15] is used to derive the VLMR. The GNSS data is available at the Nevada Geodetic Laboratory (NGL) (http://geodesy.unr.edu/NGLStationPages/GlobalStationList) [16]. Detailed information for the selected tide gauge, satellite altimetry, and GNSS data of each segment is available in [14].

Several future sea level rise scenario products have been developed to help planning and decision-making stakeholders analyze and understand vulnerabilities and future risks under scientific uncertainty. We use [17, 18] to estimate the future sea level rise rate for each segment. Sea levels under different scenarios of [17, 18] are projected to tide gauge stations and grid points, which have a resolution of 1 arc degree. If a tide gauge station exists in the segment, we use the data projected to the tide gauge station. If no tide gauge station exists in the segment, the projected grid points nearby the segment will be used. If there is more than one grid point nearby a segment, the mean value is derived and used to represent the projected sea level rise of the segment. Detailed information on the projected sea level rise data for each segment is available in [14]. In this paper, we consider the projected sea level rise under extreme scenario for 2050. For segments with tide gauge stations, the tide gauge data are integrated with the projected sea level rise data to obtain the future sea level rise rate. For segments without tide gauge stations, the combined satellite altimetry and GNSS data are integrated with the projected sea level rise data to obtain the future sea level rise rate.

After deriving the historical and future sea level rise rates, we rank them according to the percentile of the observed maximum rates, respectively. If a value falls within the highest 80 to 100th percentile, it is ranked 5 (very high). Similarly, values falling within the 60 to 80th percentile are ranked 4 (high), 40 to 60th percentile are ranked 3 (moderate), 20 to 40th percentile are ranked 2 (low), and 0 to 20th percentile are ranked 1 (very low).

2.3 Maximum annually recurring waves

Due to the sparse distribution of buoy stations in the Hawaiian Islands region, there is not enough coverage to provide wave information at a local level, i.e., for each milepost. Therefore, we use modeled wave output downloaded from Pacific Islands Ocean Observing System (PacIOOS) [19] to understand the wave conditions at each milepost. PacIOOS provides 5-day hourly wave forecasts that are calibrated using local wave buoys for the Hawaiian Islands region. Wave forecasts are simulated using WaveWatch III (WW3), surrounding the main Hawaiian Islands at an approximate resolution of 0.05 degrees, and the SWAN model, surrounding each main island at an approximate resolution of 0.31 mile (500 m) [19]. In this study, we use the wave forecasts simulated by the SWAN model, which has a finer resolution. The time span of wave data for each island varies, i.e., Oahu: 2010–2019, Maui: 2016–2019, Molokai: 2016–2019, Kauai: 2010–2019, Hawaii: 2016–2019. For each milepost, a 'virtual buoy', that is, the closest point offshore and perpendicular to the road at each milepost, is selected to obtain wave data. In this study, significant wave height was used, which is estimated as four times the square root to the zeroth order moment of the wave spectrum [19].

We extract the maximum annually recurring wave information using the method presented in [20, 21]. The process of deriving maximum annually recurring wave information is as follows. First, we identify the local peaks from the time series of significant wave heights with a time interval greater than 24 hours. Second, the peaks are divided into different bins according to incoming directions. Here, we select a 30-degree bin window, which shifts by 15-degree increments. Therefore, a maximum of 24 bins can be obtained, and there are overlaps between bins. Third, we select the three highest significant wave heights from each year and perform the generalized extreme value (GEV) fit for each bin. Then, the maximum annually recurring significant wave height (MARSWH) for each bin are derived. Finally, the wave information triplet with maximum MARSWH among all bins is selected as the annually recurring maximum wave information. We repeat this process to obtain wave information at each milepost.

After deriving the wave information triplet for each milepost, we rank the two index variables, MARSWH and corresponding peak period, according to the percentile of the observed maximum value, respectively. If a value falls within the highest 80 to 100th percentile, it is ranked 5 (very high). Similarly, values falling within the 60 to 80th percentile is ranked 4 (high), 40 to 60th percentile is ranked 3 (moderate), 20 to 40th percentile is ranked 2 (low), and 0 to 20th percentile is ranked 1 (very low).

2.4 Shoreline change

Erosion and weakening shorelines are a direct threat to coastal roads and infrastructure. Through the course of this study, we have observed both damages and an increased failure potential of nearshore state roads induced by coastal erosion.

Seasonal and storm-driven shifts in the directional transportation of sand, as well as the projected effects of sea level rise (SLR), limit the long-term numerical modeling of Hawaiian shoreline evolution. To assess the potential impact of an acceleration of shoreline change in response to rising sea levels, we interpret relative rates of shoreline change from erosion exposure forecasts developed by [20]. In [22], they describe the probabilistic method by which erosion exposure areas are determined. In [22], they use an equation for shoreline change similar to that of [23], while substituting in the geometric sediment transport model for shoreline equilibrium proposed by [24], to forecast the evolution of sandy shores on the islands of Oahu, Maui, and Kauai. Hindcast and study area limits for the model in [20] are identified from historical shorelines produced by [25]. Hindcast timespans vary between islands and study areas. Complete hindcast timespans for each island are: 1910–2007 on Oahu, 1899–2007 on Maui, and 1926–2008 on Kauai [25]. Acceleration of SLR used by [22] are taken from the Intergovernmental Panel on Climate Change (IPCC) 2013 report, AR5 high-end representative concentration pathway (RCP) 8.5 scenario – the "business as usual" scenario [26].

Shoreline change is shown in ArcGIS by digitizing the nearshore vegetation line over different periods [20]. Digitized vegetation lines (polylines), which we refer to as "Shoreline Vegetation Lines (SVLs)", are determined in [20] as the 80th percentile of the probability density function for change due to SLR of the present SVL defined during a 2006–2008 study. Projected shoreline change rates (ft/yr) are determined by dividing the length between the SVLs at the milepost, from the present vegetation line to future projected vegetation lines for SLR of 0.5, 1.1, 2.0, and 3.2 feet, by the number of years within the respective period. We assess the shoreline change at each milepost along a new polyline perpendicular to the road and extending through the SVLs, which we identify as the "measurement axis". Projected occurrence for SLR of 0.5, 1.1, 2.0, and 3.2 feet is identified by [20] using the IPCC 2013 report AR5 RCP 8.5 scenario, for the years 2030, 2050, 2075, and 2100, respectively [26]. We average the rates of shoreline erosion and accretion at each milepost over the four time periods (i.e. 2030, 2050, 2075, 2100).

Rates of interpreted averaged shoreline change are ranked into five classes according to their percentile ranges, from no change and accretion to the maximum observed averaged rate. Erosion values roughly within the highest 80 to100th percentile, are ranked 5 (very high). Similarly, erosion rates falling near the 60 to 80th percentile are ranked 4 (high), the 40 to 60th percentile is ranked 3 (moderate), and the 20 to 40th percentile is ranked 2 (low). Shoreline change values representing accretion or no change, fall roughly within the 0 to 20th percentile of maximum observed values are ranked 1 (very low). Mileposts outside of [20] are ranked based on armoring observations made in CRESI [6]. Mileposts with shoreline change values of N/A and hard armoring, where the CRESI armor ranking is greater than 3, are ranked 2 (low). Mileposts with shoreline change values of N/A and no armoring, where CRESI armor ranking [6] is less than or equal to 3, are ranked 3 (moderate).

2.5 Tsunamis

Tsunami, which is commonly caused by an earthquake in subduction zones, is one of the most devastating coastal hazards. The Hawaiian Islands region, located in the center of the Pacific Ocean, is circled by the 'Ring of Fire', a region of subduction zone volcanism. Therefore, the Hawaiian Islands region is significantly threatened by tsunamis, which result from earthquakes along the 'Ring of Fire' [27, 28]. For this reason, we take into account tsunami hazard in our assessment. We use modeled tsunami flow depth data, provided by [29], to create inundation for each milepost which in turn helps us understand how tsunami hazards affect the coastal roads in the Hawaiian Islands. The term tsunami flow depth refers to the height of tsunami water surface above ground, which can be derived by subtracting ground elevation from tsunami water level. In this study, we use two types of tsunami flow depth data: one is modeled according to historical earthquake events, and the other is based on hypothetical earthquake events. Both types of data were simulated using the model Non-hydrostatic Evolution of Ocean Wave (NEOWAVE), which is a community model developed and maintained at the University of Hawaii [30, 31]. Historical tsunami scenarios are based on the five most destructive far-field or trans-Pacific tsunamis, which were generated by the

1946 Aleutian, the 1952 Kamchatkan, the 1957 Aleutian, the 1960 Chilean, and the 1964 Alaskan earthquakes. NEOWAVE model parameters are calibrated by comparing results with well-documented runup records for those tsunamis on Hawaii shores [32–35]. The NEOWAVE model applied nested grids with increasing resolution, from 2 arcminutes (~2.3 miles) for open ocean to 0.3 arcseconds (~29.53 ft) for coastlines [32–35]. Hypothetical tsunami scenarios are based on two extreme tsunamis which apply the seismic source parameters of two hypothetical great Aleutian earthquakes. Tectonic parameters of the two great Aleutian earthquakes, with moment magnitudes of (Mw) 9.3 and 9.6, are compiled by NOAA Pacific Marine Environmental Laboratory (PMEL) and both hypothetical earthquakes are identified by a seismological study as potential sources of devastating tsunamis to Hawaii [27–29]. The model also applies nested grids with increasing resolution from 2 arcminutes (~2.3 miles) for open ocean to 0.3 arcseconds (~29.53 ft) for coastlines [29].

We use the Geographical Information System (GIS) software ArcGIS to create tsunami inundation and extract tsunami flow depth values for each milepost. Tsunami flow depths are ranked for each milepost as follows. First, mileposts are classified into three categories: Category 1 has values in the historical scenario, Category 2 has no values in the historical scenario, but has values in the hypothetical scenario, and Category 3 has no values in both historical and hypothetical scenarios. For Category 1, if a value falls within the highest 67 to 100th percentile of the observed maximum value in the historical scenario, it is ranked 5 (very high). Similarly, if a value falls within the 33 to 67th percentile, it is ranked 4 (high), and within the 0 to 33rd percentile, it is ranked 3 (moderate). For Category 2, because the tsunami flow depth in the hypothetical scenario for milepost 6 (MP 6) on Route 83, North Shore, Oahu exceeds three standard deviations of the mean, we rank it 2 and remove it from the list when searching the maximum value of Category 2. Therefore, if a value falls within the highest 50 to 100th percentile of the observed maximum value in the hypothetical scenario, it is ranked 2 (low), within the 0 to 50th percentile, it is ranked 1 (very low). All mileposts in Category 3 are ranked 1 (very low).

2.6 Storm surge

Predicting and preparing for hurricanes is a top priority for the residents and city managers of Hawaii. To assess the "worst case scenario" of inundation from storm surge, we utilize the most recent national storm surge hazard maps produced by the Storm Surge Unit (SSU) of the National Hurricane Center (NHC), National Oceanic and Atmospheric Administration (NOAA) [36].

Version 1 of the national storm surge hazard maps are published by [36] and include inundation model results for flooding caused by storm surge along the East and Gulf Coasts of the United States. Version 2, also by [36], became available in November 2018 and includes storm surge inundation estimates for the U.S. Virgin Islands, Hawaii, and Hispaniola. Measures of storm surge inundation height reflect the extents of flooding caused by storm driven uplift of the ocean surface. Estimates of storm surge inundation in this assessment are based on GIS datasets obtained through personal communication with members of the SSU and NOAA affiliates. Internal SSU issues, beyond the control of our team, have prevented a complete handover and description of the Hawaii storm surge data. As a result of the incomplete handover, there are minor errors in the projection of the data, as well as a limited understanding of the model hindcast. However, despite the shortcomings, the data still remains the best and most complete storm surge inundation data for the Hawaiian Islands. Storm surge hazard data presents hypothetical inundations found using a composite deterministic and probabilistic approach with the Sea, Lake and Overland Surges from Hurricanes (SLOSH) numerical model, developed by the National Weather Service (NWS). In the Hawaiian Islands, where steep offshore bathymetry can produce an increase in mean water level due to wave dissipation, or wave setup, the SLOSH model is loosely coupled to the third generation of the SWAN model to account for storm-related increases in mean water levels. SLOSH model forecasts consider historical atmospheric and hurricane track data, to produce a model of the wind field which drives hypothetical storm surge. However, as we mention, internal SSU issues prevents us from describing the time period for the historical atmospheric data, as well as the number and distribution of historical storm tracks.

Hawaii SLOSH model estimates include inundation scenarios for category 1 through 4 hurricanes and a broad range of storm tracks and landfall locations, consisting of hundreds of thousands of hypothetical hurricanes. Assessed storm surge inundation heights are determined as the maximum of the maximum envelops of water (MOMs), relative to a DEM of Hawaii from NOAA Office for Coastal Management (OCM) high-resolution raster elevation datasets. DEMs for each island are reoriented and divided to optimize SLOSH operation, resulting in polar oriented cells of various sizes, as small as roughly 24 ft (9 m), on each side. Within each cell, MOM values are determined in feet as a combination of all simulated inundation scenarios, with the MOM identifying the greatest observed inundation height from all simulations. Milepost assessments of storm surge inundation are sampled from the individual category of storm surge datasets, within a circular buffer centered on the milepost with a radius of 82 ft (25 m). Ranked values of storm surge inundation are determined as the percent coverage-area-weighted mean of the MOM values within the milepost buffer area. Percent coverage for each milepost buffer area is determined by first using the ArcGIS zonal statistics tool to find the buffer area overlapping with the storm surge dataset. Then, the inundation, or overlapping of the buffer area, is divided by the known total buffer area of roughly 21,000 square ft, to determine the percentage of the buffer inundated. Mean inundation height within the milepost buffer areas is also determined using the ArcGIS zonal statistics tool. Ranked values of storm surge inundation are finally calculated as the mean inundation height multiplied by the percent coverage.

Percent coverage-area-weighted mean storm surge inundation heights are ranked based on their observed distribution within the maximum observed value of each category of storm, respectively. Mileposts with inundation heights within the 50 to 100th percentile of Category 1 storm surge are ranked 5 (very high). Inundation heights greater than zero and within the 0 to 50th percentile of Category 1 storm surge, as well as the 50 to 100th percentile of Category 2 storm surge, are ranked 4 (high). If milepost inundation heights for Category 2 storm surge are greater than zero and within the 0 to 50th percentile, or within the 50 to 100th percentile for Category 3 storm surge, they are ranked 3 (moderate). Storm surge inundation heights within the 50 to 100th percentile for Category 4 storm surge, or within the 0 to 50th percentile for Category 3 storm surge are ranked 2 (low). Milepost assessments with no inundation, or with inundation heights in the 0 to 50th percentile for Category 4 storm surge are ranked 1 (very low).

3. Results: projected vulnerability for coastal highways

There are twelve regions in the State of Hawaii where coastal roads, are owned by the State, and selected due to their location to shoreline, elevation and road condition from previous ocean hazards. Of these twelve regions, four are on Oahu, two are on Molokai, three are on Maui, three are on Kauai, and one is on Hawaii. Oahu includes Waianae Coast (WC), North Shore (NS), East Shore (ES), and East Oahu (EO). Molokai includes Molokai West (KW) and Molokai East (KE). Maui includes West Maui (WM), East Maui (EM), and Central Maui (CM). Kauai includes West Kauai (W), North Kauai (N), and East Kauai (E). Hawaii includes Hilo (HILO).

Here, we present our results and how five ocean hazards: sea level rise, waves, shoreline change, tsunami, and storm surge are collectively used to rank the vulnerability of coastal highways in the State of Hawaii. A list of ocean hazards data and their associated references (superscripted) used for the Ocean Hazards Classification Scheme (OHCS) in Eq. (1) is shown in **Table 1**.

Table 2 is the Ocean Hazards Classification Scheme (OHCS) for historical and projected ocean hazards developed from the Ocean Hazards Database (OHD) [4] for state coastal roads in the State of Hawaii. In the first column is the vulnerability rank, 1 to 5, where 1 is low vulnerability and 5 is high vulnerability. The remaining columns are the associated Variables and their resulting rates, heights or depths according to the methodology described in Section 2, using 302 mileposts across the State from [2]. Using **Table 2**, we rank each Variable (1 to 5) and apply it to Eq. (1) to retrieve the OHCS ranking, that is, a combined ranking of vulnerability for sea level rise, significant wave height, shoreline change, tsunami and storm surge. Our results are listed as follows.

Oahu Waianae Coast (WC), **Figures 2-4**: Includes 39 mileposts. The OHCS vulnerability ranking ranges from 1 to 3, with a few higher ranking outliers of 5, 5, and 6 at MPs 19 + 0.55, 16 + 0.41 and 10 + 0.25, respectively. In this region, sea level rise ranges from 1 to 2, significant wave height ranges from 1 to 2, shoreline change ranges from 2 to 3, tsunami ranges from 1 to 4, and storm surge ranges from 1 to 4. The outliers, i.e. the OHCS rankings of 5 and 6 at MP 19 + 0.55, 16 + 0.41 and 10 + 0.25, are a result from the increased tsunami and storm surge rankings, due to proximity and elevation of the road to the shoreline at those particular locations.

Oahu North Shore (NS), **Figure 5**: Includes 19 mileposts. The OHCS vulnerability ranking ranges from 2 to 15. In this region, sea level rise is 2, significant wave height ranges from 1 to 3, shoreline change ranges from 2 to 3, tsunami ranges from 1 to 4, and storm surge is 1 with a three MPs ranked at 4. Although most of the OHCS values range from 7 or below, the three MPs worth noting, i.e. MP 3 + 0.66, 4 + 0.49 and 6, with a ranking of 15, 15, and 10, respectively, are the MPs with a storm surge ranking of 4, compared to the other MPs with a storm surge ranking of 1.

Variable	Classification	Description [units]
1	Sea Level Rise	2050 Sea Level Rise Rate [9, 10, 12, 14, 17, 18] (1905–2050, extreme scenario) [in/yr]
2	Maximum Annually Recurring Waves	Significant Wave Height [19, 20] (2010–2018) [ft]
3	Shoreline Change	Mean Shoreline Change Rate [6, 20] (2008–2100) [ft/yr]
4	Tsunami	Inundation Depth (Historical and Hypothetical) [29, 32–35] [ft]
5	Storm Surge	Category 1–4 Storm Inundation Depth [36] (Hypothetical) [ft]

Table 1.
Historical and projected ocean hazards variables used in the Ocean Hazards Classification Scheme (OHCS) for State coastal roads in the State of Hawaii. For more detailed explanation of each, refer to [4]. 12 inches = 1 foot = 0.3048 meters.

Vunerability Rank	Variable 1	Variable 2	Variable 3	Variable 4	Variable 5
	Sea Level Rise	**Maximum Annually Recurring Waves**	**Shoreline Change**	**Tsunami**	**Storm Surge**
	2050 Sea Level Rise Rate [9, 10, 12, 14, 17, 18] (1905–2050, extreme scenario)	**Significant Wave Height [19, 20] (2010–2018)**	**Mean Shoreline Change Rate [6, 20] (2008–2100)**	**Tsunami Inundation [29, 32–35] (Historical and Hypothetical)**	**Weighted Mean Storm Surge Inundation [36] (Hypothetical)**
1	<0.1 in/yr	<7 ft	<0 ft/yr	No inundation or Hypothetical inundation <16 ft with no Historical Inundation	No Inundation or Category 4 Inundation <4 ft
2	0.1 to 0.2 in/yr	7 to 14 ft	0 to 2 ft/yr & "N/A" with >3 Armoring Ranking	Hypothetical inundation ≥16 ft with no Historical Inundation	Category 3 Inundation <4 ft or Category 4 Inundation of 4 to 8 ft
3	0.2 to 0.3 in/yr	14 to 21 ft	2 to 5 ft/yr & "N/A" with ≤3 Armoring Ranking	Historical inundation <6 ft	Category 3 Inundation of 4 to 7 ft or Category 2 Inundation <1 ft
4	0.3 to 0.4 in/yr	21 to 29 ft	5 to 7 ft/yr	Historical inundation of 6 to 12 ft	Category 2 Inundation of 1 to 6 ft or Category 1 Inundation <1 ft
5	> 0.4 in/yr	> 29 ft	> 7 ft/yr	Historical inundation ≥12 ft	Category 1 Inundation of 1 to 4 ft

Table 2.
Ocean Hazards Classification Scheme (OHCS) for historical and projected ocean hazards developed from [34] for State coastal roads in the State of Hawaii. 12 inches = 1 foot = 0.3048 meters.

Oahu East Shore (ES), **Figures 6-8**: Includes 44 mileposts. The OHCS vulnerability ranking ranges from 1 to 12. In this region, sea level rise is 2, significant wave height ranges from 1 to 2, shoreline change ranges from 1 to 3, tsunami ranges from 1 to 4, and storm surge ranges from 1 to 5. We see particularly high OHCS rankings of 9 to 12 at certain MPs. These regions with OHCS values of 9 to 12, is a result from the increased tsunami and storm surge rankings.

Oahu East Oahu (EO), **Figures 9** and **10**: Includes 20 mileposts. The OHCS vulnerability ranking ranges from 1 to 10. In this region, sea level rise ranges from 1 to 2, significant wave height ranges from 1 to 2, shoreline change ranges from 2 to 3, tsunami ranges from 1 to 4, and storm surge ranges from 1 to 5. High OHCS rankings of 7 to 10, is a result from the increased tsunami and storm surge rankings.

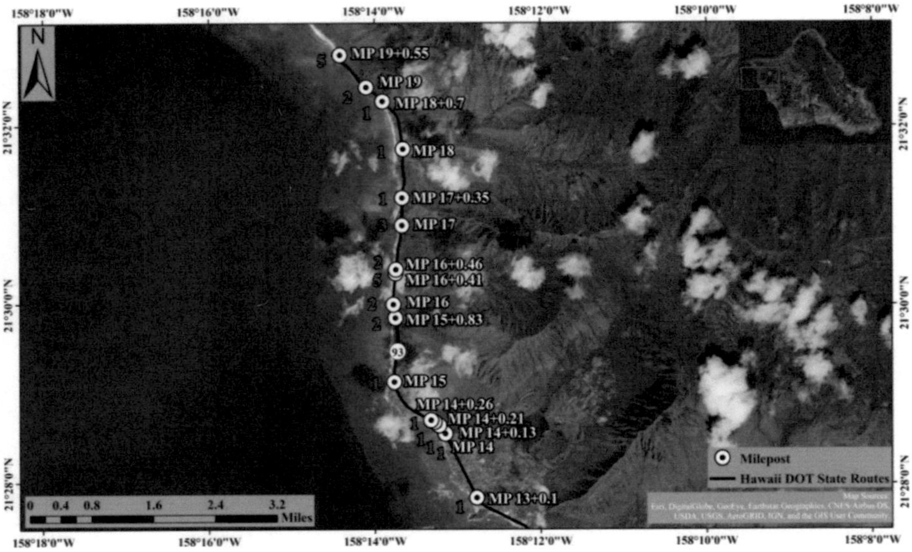

Figure 2.
Ocean Hazards Classification Scheme (OHCS) ranking for Oahu Waianae Coast (WC) MP 13 + 0.1 to 19 + 0.55. The OHCS consists of five variables: (i) sea level rise 1905–2050, (ii) maximum annually recurring significant wave height 2010–2018, (iii) shoreline change 2008–2100 and CRESI Armoring, (iv) historical and hypothetical tsunami, and (v) Category 1,2,3,4 hypothetical storm surge. Rankings of ocean hazard increase from 1 to a theoretical maximum of 100.

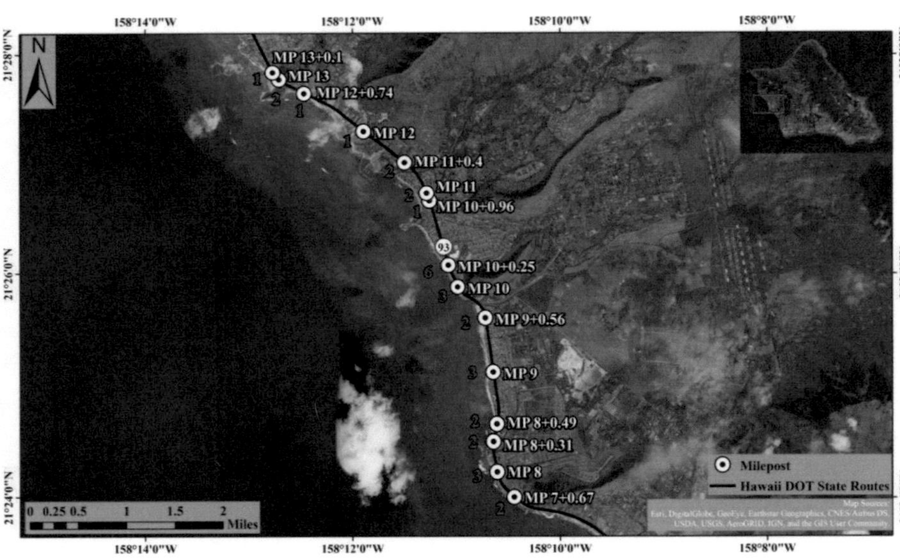

Figure 3.
Ocean Hazards Classification Scheme (OHCS) ranking for Oahu Waianae Coast (WC) MP 7 + 0.67_13 + 0.1. The OHCS consists of five variables: (i) sea level rise 1905–2050, (ii) maximum annually recurring significant wave height 2010–2018, (iii) shoreline change 2008–2100 and CRESI Armoring, (iv) historical and hypothetical tsunami, and (v) Category 1,2,3,4 hypothetical storm surge. Rankings of ocean hazard increase from 1 to a theoretical maximum of 100.

Molokai Molokai West (KW), **Figure 11**: Includes 5 mileposts. The OHCS vulnerability ranking ranges from 13 to 17, with a low OHCS ranking outlier of 3 at MP 2. In this region for OHCS rankings of 13 to 17, the sea level rise is 5, significant wave height is 1, shoreline change ranges from 2 to 3, tsunami is 3, and storm surge

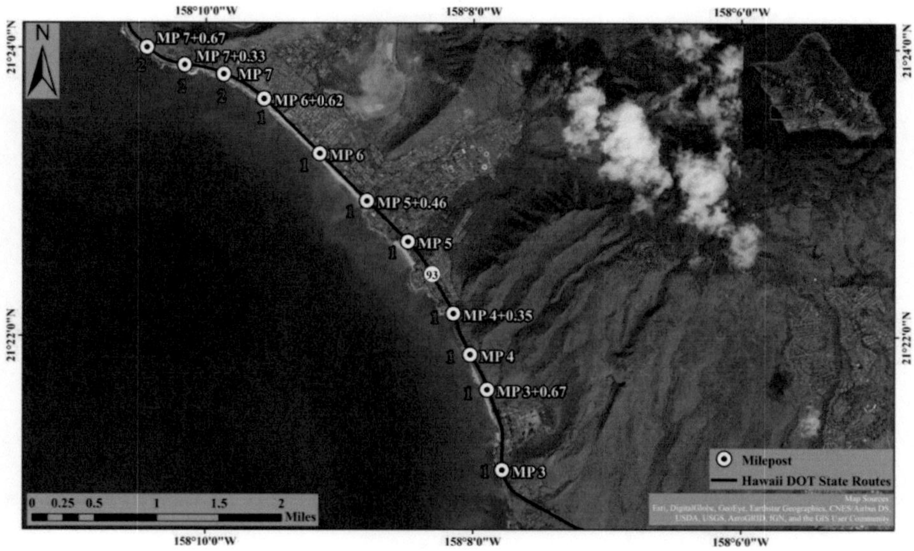

Figure 4.
Ocean Hazards Classification Scheme (OHCS) ranking for Oahu Waianae Coast (WC) MP 3 to 7 + 0.67. The OHCS consists of five variables: (i) sea level rise 1905–2050, (ii) maximum annually recurring significant wave height 2010–2018, (iii) shoreline change 2008–2100 and CRESI Armoring, (iv) historical and hypothetical tsunami, and (v) Category 1,2,3,4 hypothetical storm surge. Rankings of ocean hazard increase from 1 to a theoretical maximum of 100.

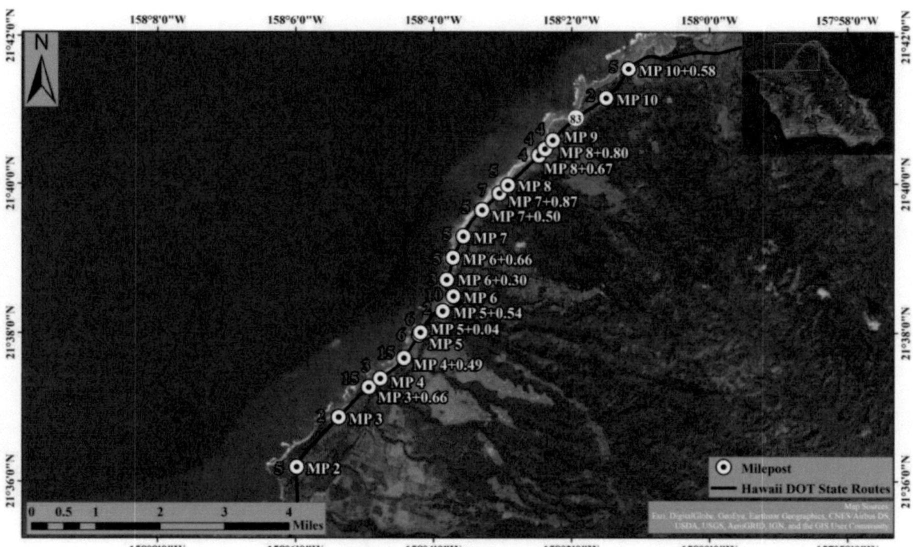

Figure 5.
Ocean Hazards Classification Scheme (OHCS) ranking for Oahu North Shore (NS) MP 2 to 10 + 0.58. The OHCS consists of five variables: (i) sea level rise 1905–2050, (ii) maximum annually recurring significant wave height 2010–2018, (iii) shoreline change 2008–2100 and CRESI Armoring, (iv) historical and hypothetical tsunami, and (v) Category 1,2,3,4 hypothetical storm surge. Rankings of ocean hazard increase from 1 to a theoretical maximum of 100.

ranges from 4 to 5. High OHCS rankings of 13 to 17 is a result of higher rankings for sea level rise, storm surge and tsunami inundation in this region.

Molokai Molokai East (KE), **Figures 12-14**: Includes 49 mileposts. The OHCS vulnerability ranking ranges from 3 to 33. In this region, the sea level rise is 5, significant wave height ranges from 1 to 2, shoreline change is 3, tsunami ranges from

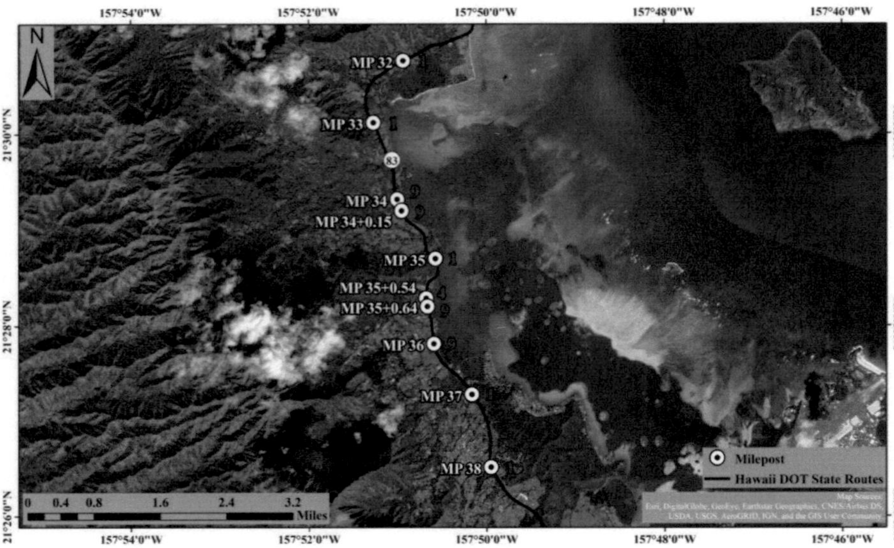

Figure 6.
Ocean Hazards Classification Scheme (OHCS) ranking for Oahu East Shore (ES) MP 32 to 38. The OHCS consists of five variables: (i) sea level rise 1905–2050, (ii) maximum annually recurring significant wave height 2010–2018, (iii) shoreline change 2008–2100 and CRESI Armoring, (iv) historical and hypothetical tsunami, and (v) Category 1,2,3,4 hypothetical storm surge. Rankings of ocean hazard increase from 1 to a theoretical maximum of 100.

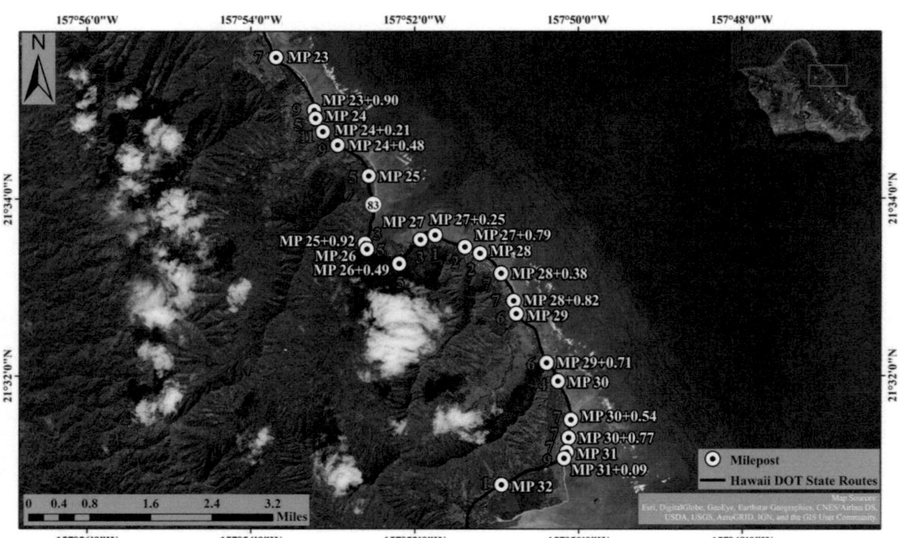

Figure 7.
Ocean Hazards Classification Scheme (OHCS) ranking for Oahu East Shore (ES) MP 23 to 32. The OHCS consists of five variables: (i) sea level rise 1905–2050, (ii) maximum annually recurring significant wave height 2010–2018, (iii) shoreline change 2008–2100 and CRESI Armoring, (iv) historical and hypothetical tsunami, and (v) Category 1,2,3,4 hypothetical storm surge. Rankings of ocean hazard increase from 1 to a theoretical maximum of 100.

1 to 5, and storm surge ranges from 1 to 5. High OHCS rankings is a result of high rankings for sea level rise, storm surge and tsunami inundation in this region.

Maui West Maui (WM), **Figures 15-17**: Includes 48 mileposts. The OHCS vulnerability ranking ranges from 1 to 14. In this region, the sea level rise is 2, significant wave height ranges from 1 to 2, shoreline change ranges from 1 to 3, tsunami

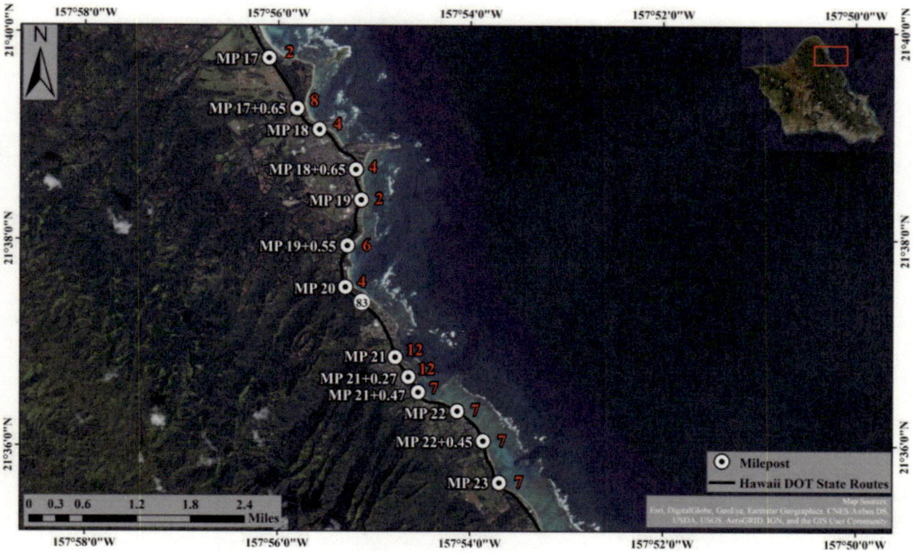

Figure 8.
Ocean Hazards Classification Scheme (OHCS) ranking for Oahu East Shore (ES) MP 17 to 23. The OHCS consists of five variables: (i) sea level rise 1905–2050, (ii) maximum annually recurring significant wave height 2010–2018, (iii) shoreline change 2008–2100 and CRESI Armoring, (iv) historical and hypothetical tsunami, and (v) Category 1,2,3,4 hypothetical storm surge. Rankings of ocean hazard increase from 1 to a theoretical maximum of 100.

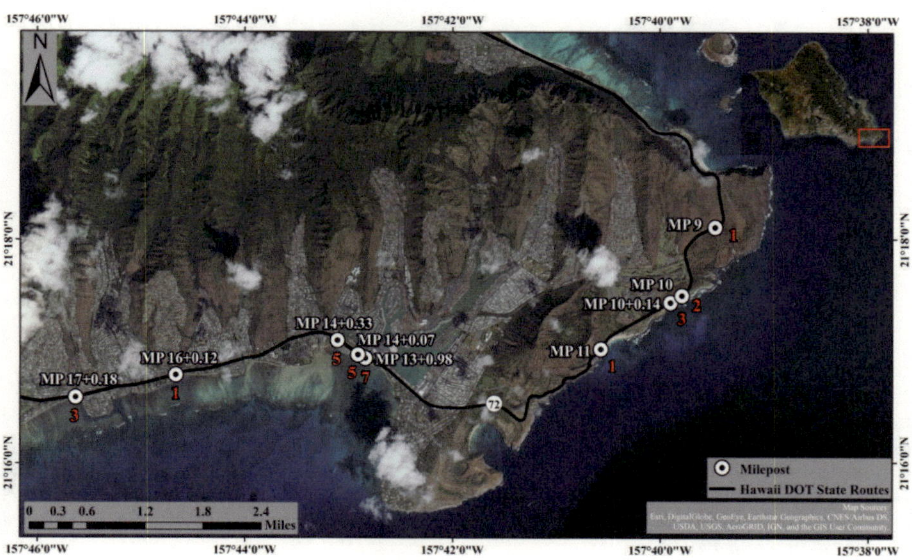

Figure 9.
Ocean Hazards Classification Scheme (OHCS) ranking for Oahu East Oahu (EO) MP 9 to 17 + 0.18. The OHCS consists of five variables: (i) sea level rise 1905–2050, (ii) maximum annually recurring significant wave height 2010–2018, (iii) shoreline change 2008–2100 and CRESI Armoring, (iv) historical and hypothetical tsunami, and (v) Category 1,2,3,4 hypothetical storm surge. Rankings of ocean hazard increase from 1 to a theoretical maximum of 100.

ranges from 1 to 5, and storm surge ranges from 1 to 5. High OHCS rankings is a result of high rankings for storm surge and tsunami inundation in this region.

Maui East Maui (EM), **Figure 18**: Includes 11 mileposts. The OHCS vulnerability ranking ranges from 6 to 10. In this region, the sea level rise is 2, significant wave height is 1, shoreline change ranges from 2 to 3, tsunami ranges from 4 to 5, and

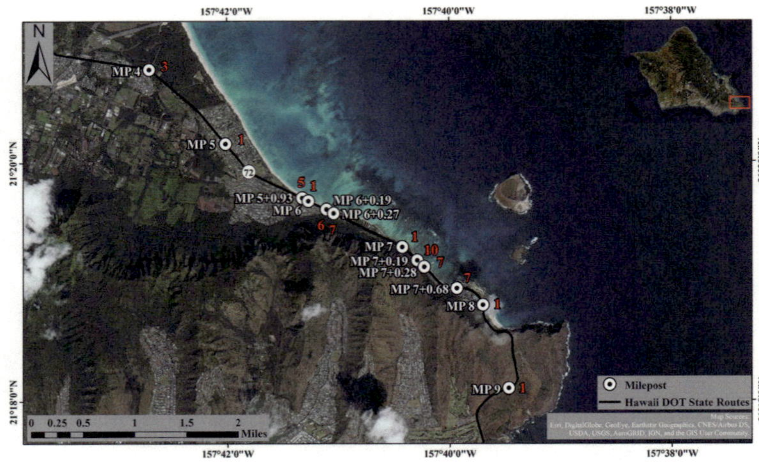

Figure 10.
Ocean Hazards Classification Scheme (OHCS) ranking for Oahu East Oahu (EO) MP 4 to 9. The OHCS consists of five variables: (i) sea level rise 1905–2050, (ii) maximum annually recurring significant wave height 2010–2018, (iii) shoreline change 2008–2100 and CRESI Armoring, (iv) historical and hypothetical tsunami, and (v) Category 1,2,3,4 hypothetical storm surge. Rankings of ocean hazard increase from 1 to a theoretical maximum of 100.

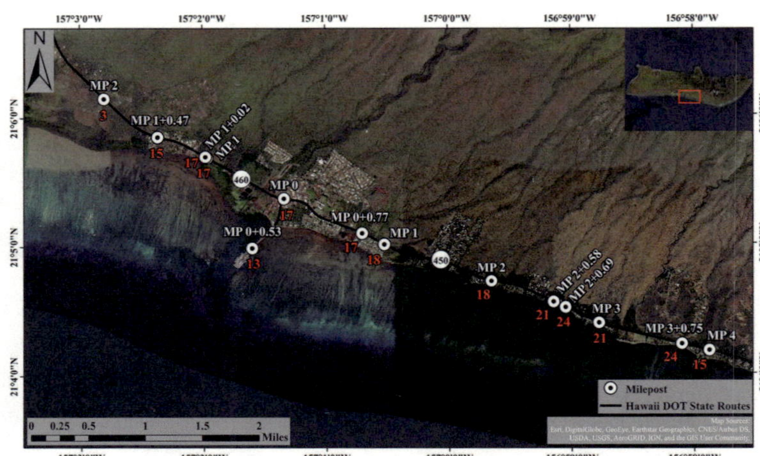

Figure 11.
Ocean Hazards Classification Scheme (OHCS) ranking for Molokai Molokai West (KW) MP 2 to East 4. The OHCS consists of five variables: (i) sea level rise 1905–2050, (ii) maximum annually recurring significant wave height 2010–2018, (iii) shoreline change 2008–2100 and CRESI Armoring, (iv) historical and hypothetical tsunami, and (v) Category 1,2,3,4 hypothetical storm surge. Rankings of ocean hazard increase from 1 to a theoretical maximum of 100.

storm surge ranges from 2 to 5. High OHCS rankings is a result of high rankings for storm surge and tsunami inundation in this region.

Maui Central Maui (CM), **Figures 19** and **20**: Includes 13 mileposts. The OHCS vulnerability ranking ranges from 3 to 16. In this region, the sea level rise ranges from 2 to 5, significant wave height ranges from 1 to 5, shoreline change ranges from 2 to 5, tsunami ranges from 1 to 5, and storm surge ranges from 1 to 5. High OHCS rankings is generally a result of high rankings for sea level rise, storm surge and tsunami inundation in this region. However, significant wave height contributes to high OHCS rankings at MPs 8 + 0.42 and 8 + 0.63 and shoreline change at MP 0 + 0.05.

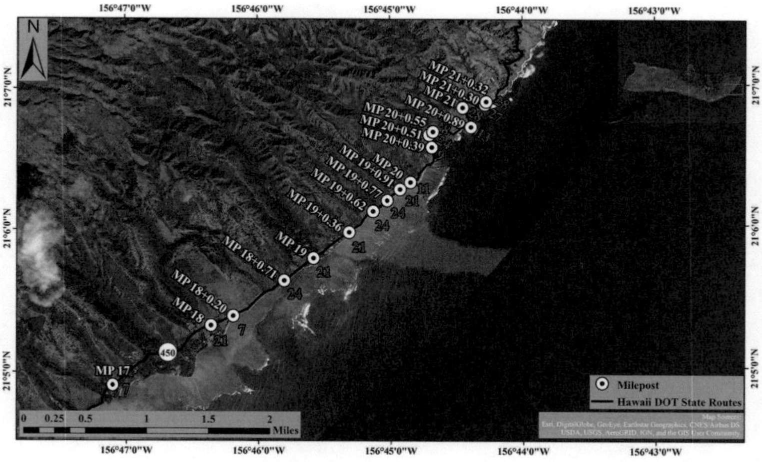

Figure 12.
Ocean Hazards Classification Scheme (OHCS) ranking for Molokai Molokai East (KE) MP 17 to 21 + 0.32. The OHCS consists of five variables: (i) sea level rise 1905–2050, (ii) maximum annually recurring significant wave height 2010–2018, (iii) shoreline change 2008–2100 and CRESI Armoring, (iv) historical and hypothetical tsunami, and (v) Category 1,2,3,4 hypothetical storm surge. Rankings of ocean hazard increase from 1 to a theoretical maximum of 100.

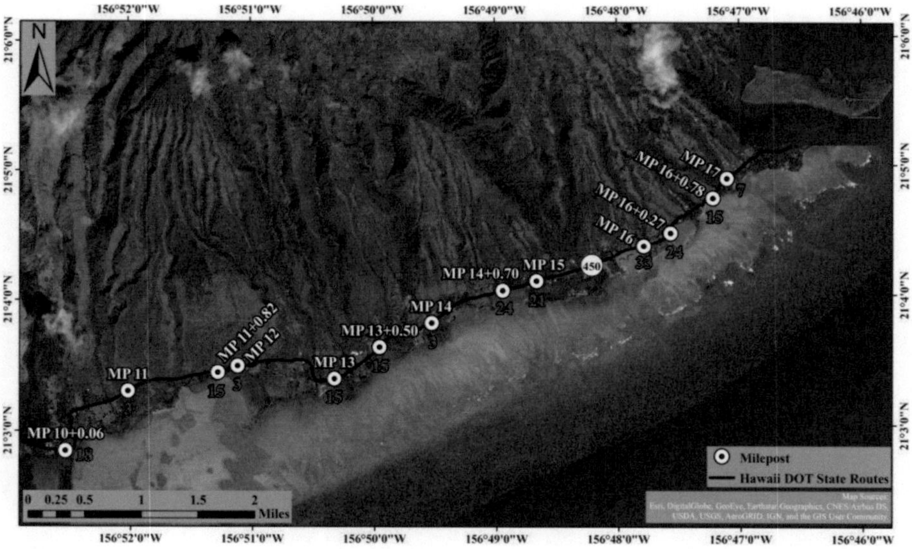

Figure 13.
Ocean Hazards Classification Scheme (OHCS) ranking for Molokai Molokai East (KE) MP 10 + 0.06 to 17. The OHCS consists of five variables: (i) sea level rise 1905–2050, (ii) maximum annually recurring significant wave height 2010–2018, (iii) shoreline change 2008–2100 and CRESI Armoring, (iv) historical and hypothetical tsunami, and (v) Category 1,2,3,4 hypothetical storm surge. Rankings of ocean hazard increase from 1 to a theoretical maximum of 100.

Kauai West Kauai (W), **Figure 21**: Includes 11 mileposts. The OHCS vulnerability ranking ranges from 4 to 11, with a low OHCS outlier of 1 at MP 24 + 0.91. In this region, the sea level rise is 2, significant wave height is 1, shoreline change ranges from 1 to 4, tsunami ranges from 3 to 4, and storm surge ranges from 1 to 5. High OHCS rankings is a result of high rankings for storm surge and tsunami inundation in this region.

Kauai North Kauai (N), **Figure 22**: Includes 8 mileposts. The OHCS vulnerability ranking ranges from 3 to 11. In this region, the sea level rise is 2, significant wave

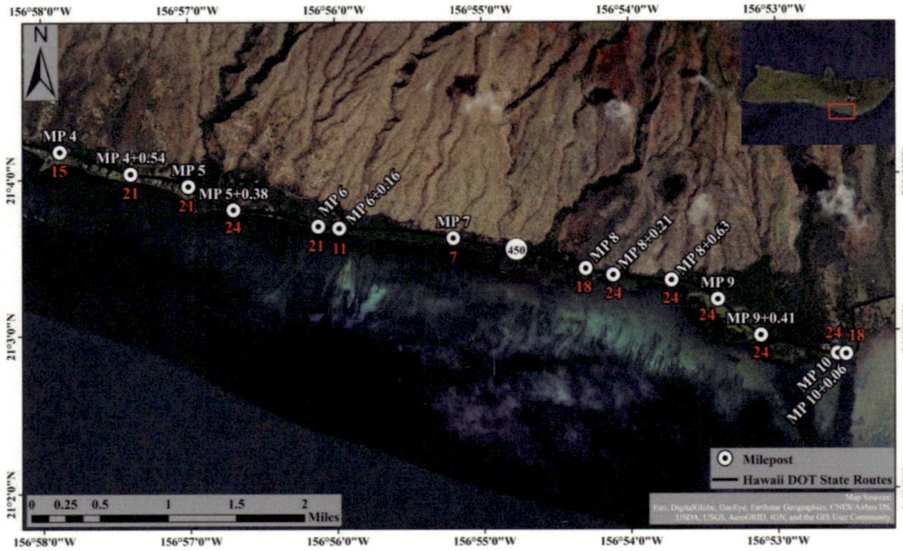

Figure 14.
Ocean Hazards Classification Scheme (OHCS) ranking for Molokai Molokai East (KE) MP 4 to 10 + 0.06. The OHCS consists of five variables: (i) sea level rise 1905–2050, (ii) maximum annually recurring significant wave height 2010–2018, (iii) shoreline change 2008–2100 and CRESI Armoring, (iv) historical and hypothetical tsunami, and (v) Category 1,2,3,4 hypothetical storm surge. Rankings of ocean hazard increase from 1 to a theoretical maximum of 100.

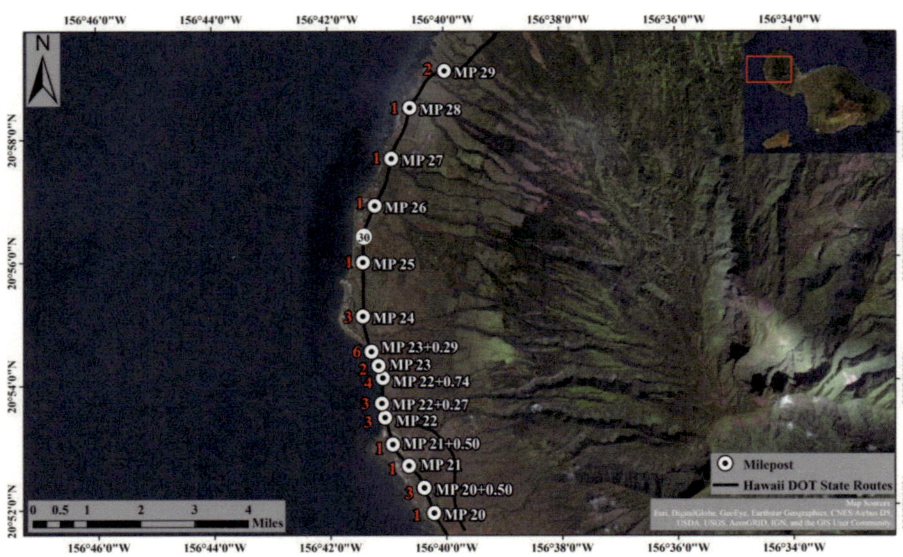

Figure 15.
Ocean Hazards Classification Scheme (OHCS) ranking for Maui West Maui (WM) MP 20 to 29. The OHCS consists of five variables: (i) sea level rise 1905–2050, (ii) maximum annually recurring significant wave height 2010–2018, (iii) shoreline change 2008–2100 and CRESI Armoring, (iv) historical and hypothetical tsunami, and (v) Category 1,2,3,4 hypothetical storm surge. Rankings of ocean hazard increase from 1 to a theoretical maximum of 100.

height ranges from 1 to 2, shoreline change ranges from 2 to 5, tsunami ranges from 3 to 4, and storm surge ranges from 1 to 5. High OHCS rankings is a result of higher rankings for storm surge and shoreline change in this region.

Kauai East Kauai (E), **Figure 23**: Includes 13 mileposts. The OHCS vulnerability ranking ranges from 2 to 9. In this region, the sea level rise is 2, significant wave

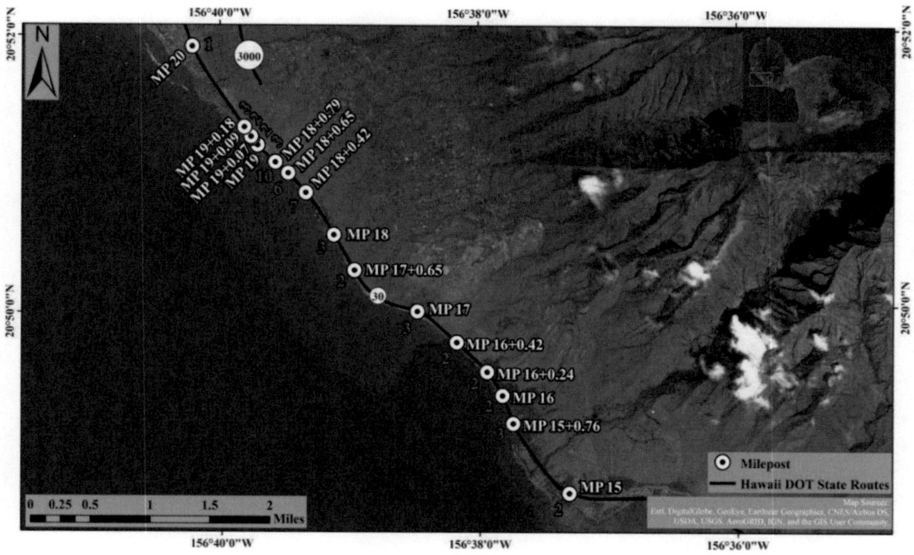

Figure 16.
Ocean Hazards Classification Scheme (OHCS) ranking for Maui West Maui (WM) MP 15 to 20. The OHCS consists of five variables: (i) sea level rise 1905–2050, (ii) maximum annually recurring significant wave height 2010–2018, (iii) shoreline change 2008–2100 and CRESI Armoring, (iv) historical and hypothetical tsunami, and (v) Category 1,2,3,4 hypothetical storm surge. Rankings of ocean hazard increase from 1 to a theoretical maximum of 100.

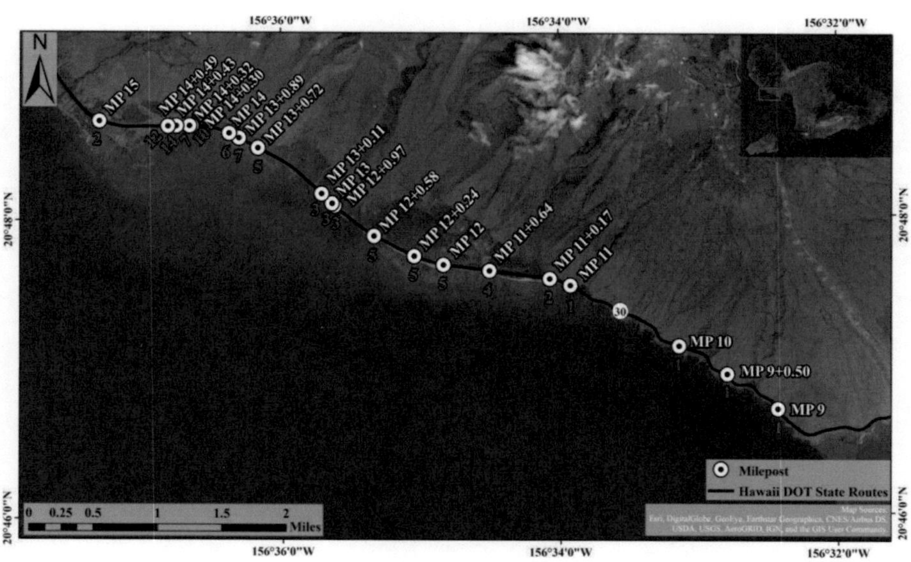

Figure 17.
Ocean Hazards Classification Scheme (OHCS) ranking for Maui West Maui (WM) MP 9 to 15. The OHCS consists of five variables: (i) sea level rise 1905–2050, (ii) maximum annually recurring significant wave height 2010–2018, (iii) shoreline change 2008–2100 and CRESI Armoring, (iv) historical and hypothetical tsunami, and (v) Category 1,2,3,4 hypothetical storm surge. Rankings of ocean hazard increase from 1 to a theoretical maximum of 100.

height ranges from 1 to 2, shoreline change ranges from 2 to 3, tsunami ranges from 1 to 4, and storm surge ranges from 1 to 5. High OHCS rankings is a result of higher rankings for shoreline change, tsunami and storm surge in this region.

Hawaii Hilo (HILO), **Figure 24**: Includes 12 mileposts. The OHCS vulnerability ranking ranges from 4 to 18. In this region, the sea level rise is 4, significant

Figure 18.
Ocean Hazards Classification Scheme (OHCS) ranking for Maui East Maui (EM) MP 1 to 3 + 0.14. The OHCS consists of five variables: (i) sea level rise 1905–2050, (ii) maximum annually recurring significant wave height 2010–2018, (iii) shoreline change 2008–2100 and CRESI Armoring, (iv) historical and hypothetical tsunami, and (v) Category 1,2,3,4 hypothetical storm surge. Rankings of ocean hazard increase from 1 to a theoretical maximum of 100.

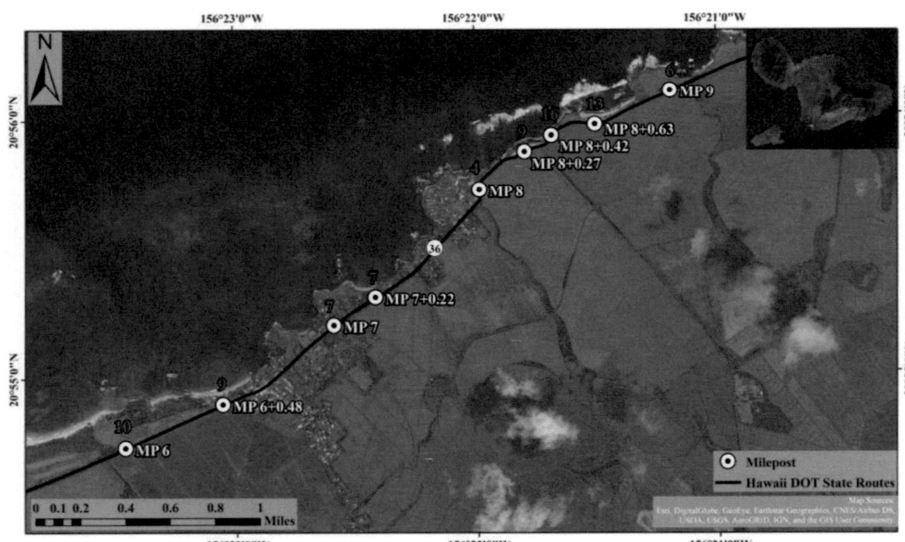

Figure 19.
Ocean Hazards Classification Scheme (OHCS) ranking for Maui Central Maui (CM) MP 6 to 9. The OHCS consists of five variables: (i) sea level rise 1905–2050, (ii) maximum annually recurring significant wave height 2010–2018, (iii) shoreline change 2008–2100 and CRESI Armoring, (iv) historical and hypothetical tsunami, and (v) Category 1,2,3,4 hypothetical storm surge. Rankings of ocean hazard increase from 1 to a theoretical maximum of 100.

wave height ranges from 1 to 2, shoreline change ranges from 2 to 3, tsunami ranges from 1 to 5, and storm surge ranges from 1 to 5. High OHCS rankings is a result of higher rankings for sea level rise, tsunami and storm surge in this region.

In summary from our results, sea level rise ranges from 1 to 5, waves ranges from 1 to 5, and the OHCS ranges from 1 to 33. Although the OHCS equation allows

Figure 20.
Ocean Hazards Classification Scheme (OHCS) ranking for Maui Central Maui (CM) MP 0 to 0 + 0.71. The OHCS consists of five variables: (i) sea level rise 1905–2050, (ii) maximum annually recurring significant wave height 2010–2018, (iii) shoreline change 2008–2100 and CRESI Armoring, (iv) historical and hypothetical tsunami, and (v) Category 1,2,3,4 hypothetical storm surge. Rankings of ocean hazard increase from 1 to a theoretical maximum of 100.

Figure 21.
Ocean Hazards Classification Scheme (OHCS) ranking for Kauai West Kauai (W) MP 24 to 28. The OHCS consists of five variables: (i) sea level rise 1905–2050, (ii) maximum annually recurring significant wave height 2010–2018, (iii) shoreline change 2008–2100 and CRESI Armoring, (iv) historical and hypothetical tsunami, and (v) Category 1,2,3,4 hypothetical storm surge. Rankings of ocean hazard increase from 1 to a theoretical maximum of 100.

a value up to 100, OHCS only went up to 33, showing that no locations have all Variables at high vulnerability (i.e. 5), but rather a one or two Variables may be at rank 5 while the other Variables remain low (i.e. 1 or 2).

Another result shows that the island of Molokai has the highest OHCS overall. The Variables that contribute to the high OHCS includes sea level rise, tsunami and storm surge, all of which were nearly ranked at 5.

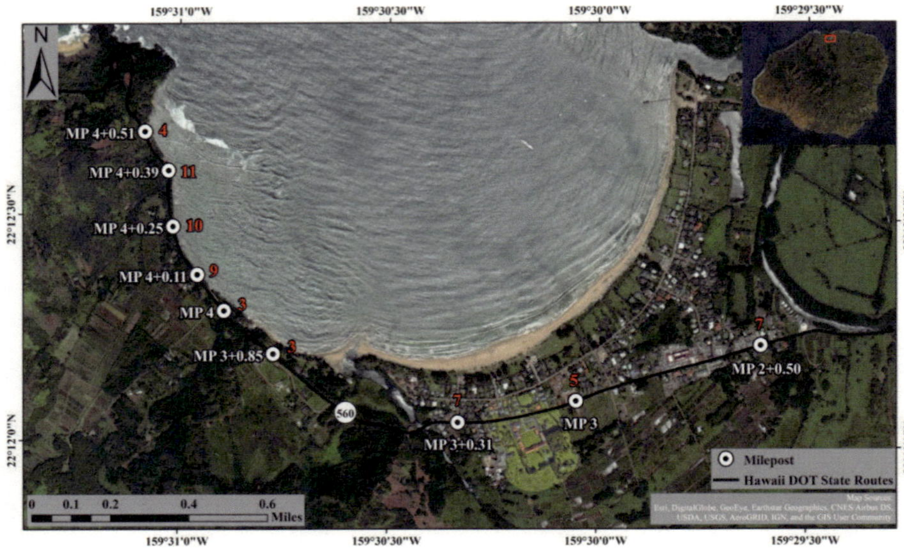

Figure 22.
Ocean Hazards Classification Scheme (OHCS) ranking for Kauai North Kauai (N) MP 2 + 0.5 to 4 + 0.51. The OHCS consists of five variables: (i) sea level rise 1905–2050, (ii) maximum annually recurring significant wave height 2010–2018, (iii) shoreline change 2008–2100 and CRESI Armoring, (iv) historical and hypothetical tsunami, and (v) Category 1,2,3,4 hypothetical storm surge. Rankings of ocean hazard increase from 1 to a theoretical maximum of 100.

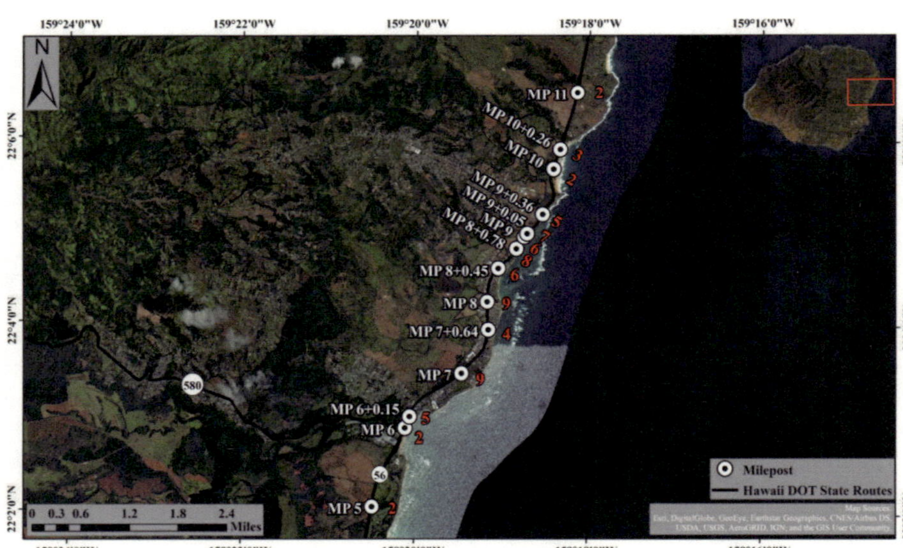

Figure 23.
Ocean Hazards Classification Scheme (OHCS) ranking for Kauai East Kauai (E) MP 5 to 11. The OHCS consists of five variables: (i) sea level rise 1905–2050, (ii) maximum annually recurring significant wave height 2010–2018, (iii) shoreline change 2008–2100 and CRESI Armoring, (iv) historical and hypothetical tsunami, and (v) Category 1,2,3,4 hypothetical storm surge. Rankings of ocean hazard increase from 1 to a theoretical maximum of 100.

A third result is that the Variable, storm surge, is consistently the largest contributor in coastal vulnerability on state roads for all islands. This is shown in the ranking of all Variables which largely show a storm surge of rank 5 at most locations, where the other Variables remain at 1 or 2. Tsunamis are the second largest contributor in our results. Although sea level rise was not one of the highest

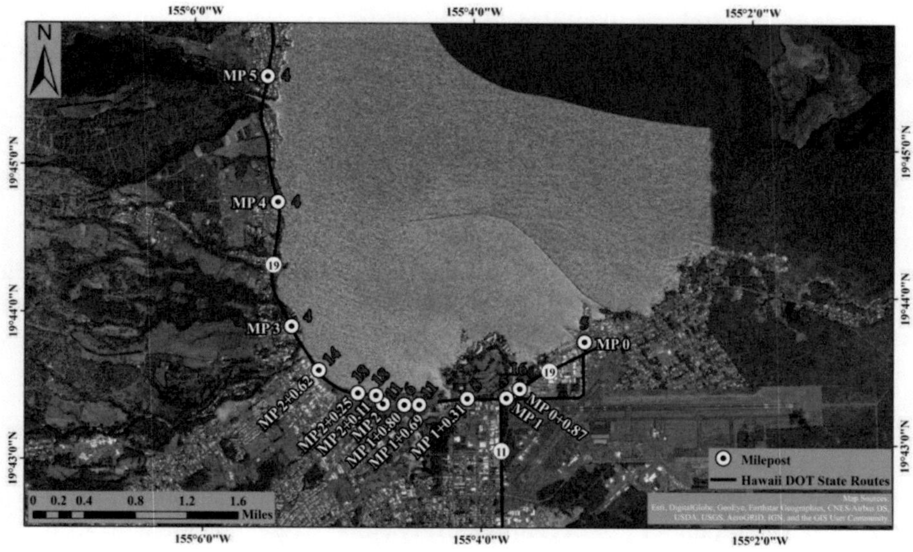

Figure 24.
Ocean Hazards Classification Scheme (OHCS) ranking for Hawaii Hilo (HILO) MP 0 to 5. The OHCS consists of five variables: (i) sea level rise 1905–2050, (ii) maximum annually recurring significant wave height 2010–2018, (iii) shoreline change 2008–2100 and CRESI Armoring, (iv) historical and hypothetical tsunami, and (v) Category 1,2,3,4 hypothetical storm surge. Rankings of ocean hazard increase from 1 to a theoretical maximum of 100.

contributors, it should be considered a main contributor since the sea level rise inundation amplifies storm surge and tsunami inundation.

4. Conclusion

The high rankings of storm surge inundation and tsunami inundation are due to lower road elevation, which puts the road at greater risk. Road relocation inland is recommended, if possible. Where road relocation is not possible, and usually not an option for state roads in Hawaii, elevating the road infrastructure (and therefore other surrounding infrastructure) should be taken into consideration in community planning and development. To reinforce the elevated road, hardening should be included also.

Although our Variables we consider: (1) sea level rise, (2) waves, (3) shoreline change, (4) tsunamis, and (5) storm surge, work for our study region, i.e. the Hawaiian Islands, one should be aware that assessing vulnerability is "location specific". This means that natural hazards affecting an area depend on many factors such as the geology, oceanic, bathymetric, and climate trends in a location. These factors differ region to region. Each coastal region should develop their own vulnerability ranking method to include or not include Variables which most likely affect their region.

While natural hazard exposure to infrastructure is important, other multiple indicators should also be considered. For roadways this may include traffic volume, population served, accessibility, connectivity, reliability, land use, and roadway connection to critical infrastructures, such as hospitals and police stations [37]. However, this type of data changes frequently as land use develops at a rapid pace or additionally roads may be added. Also adding these additional indicators may change the CVI (or OHCS).

Coastal hazard and risk not only comes in the form of the physical processes on the ecosystem or built infrastructure, but also through social perceptions, as well. Perceptions of coastal hazards and risks and community support for engineered adaptation methods are important for implementation among different stakeholder groups (experts, businesses, and community members) [38].

By understanding the vulnerability of a region, we may assign what adaptation method to use in vulnerable coastal regions dealing with climate change, in particular, inundation. These engineered adaptation methods include offshore barriers, coastal armoring, elevated development, floating development, floodable development, living shorelines, and managed retreat [39]. In the future, if we want to continue to live on coast, we must adapt.

Acknowledgements

Funding provided by the State of Hawaii Department of Transportation Highways Division, under Project Number HWY-06-16. We would like to also acknowledge our affiliations to the following at the University of Hawaii at Manoa: Department of Civil and Environmental Engineering, Sea Grant College Program, and the Coastal Hydraulics Engineering Resilience (CHER) Lab. Publications fees provided by Research & Training Revolving Fund from the Civil and Environmental Engineering Department at the University of Hawaii at Manoa.

Author details

Oceana Francis[1*], Linqiang Yang[1], Harrison Togia[1] and Gleb Panteleev[2]

1 University of Hawaii at Manoa, Honolulu, Hawaii, USA

2 Naval Research Laboratory, Stennis Space Center, Mississippi, USA

*Address all correspondence to: oceanaf@hawaii.edu

References

[1] Richmond, B.M., Fletcher, C.H., Grossman, E.E., and Gibbs, A.E. (2001). Islands at Risk: Coastal Hazard Assessment and Mapping in The Hawaiian Islands. Environmental Geosciences, 8 (1): 21-37. doi: https://doi.org/10.1046/j.1526-0984.2001.008001021.x.

[2] Francis, O., Brandes, H., Zhang, G., Ma, D., Yang, L., Doygun, O., Togia, H., Rossi, C., and Costanzo, G. (2019). State of Hawaii Statewide Coastal Highway Program Report. Prepared for the State of Hawaii Department of Transportation, Project Number HWY-06-16, August 21, 2019, https://hidot.hawaii.gov/highways/files/2019/09/State-of-Hawaii-Statewide-Coastal-Highway-Program-Report_Final_2019.pdf.

[3] Hess, K., Schmalz, R., Zervas, C., and Collier, W. (1999). Tidal constituents and residual interpolation (TCARI): A new method for the tidal correction of bathymetric data. NOAA Technical Report, NOS CS 4, 99.

[4] Francis, O., Yang, L., Togia, H., and Tumino Di Costanzo, G. (2020). Ocean Hazards Database (OHD) for the State of Hawaii Statewide Coastal Highway Program Report. Mendeley Data, v11 http://dx.doi.org/10.17632/7p3hyypmjm.11.

[5] Gornitz, V. (1991). Global coastal hazards from future sea level rise. Palaeogeography Palaeoclimatology Palaeoecology, 89, 379-398, doi:10.1016/0031-0182(91)90173-O.

[6] Brandes, H., Doygun, O., Rossi, C., Francis, O., Yang, L., and Togia, H. (2019). Coastal Road Exposure Susceptibility Index (CRESI) for the State of Hawaii Statewide Coastal Highway Program Report. Mendeley Data, v2 http://dx.doi.org/10.17632/frr3fsx3j6.2.

[7] Onat, Y., Marchant, M., Francis, O., and Kim, K. (2018). Coastal exposure of the Hawaiian Islands using GIS-based index modeling. Ocean & Coastal Management, 163, 113-129, ISSN 0964-5691, https://doi.org/10.1016/j.ocecoaman.2018.06.003.

[8] Onat, Y., Francis, O., and Kim, K. (2018). Vulnerability assessment and adaptation to sea level rise in high-wave environments: A case study on O'ahu, Hawai'i, Ocean & Coastal Management, 157, 147-159, ISSN 0964-5691, https://doi.org/10.1016/j.ocecoaman.2018.02.021.

[9] Holgate, S.J., Matthews, A., Woodworth, P.L., Rickards, L.J., Tamisiea, M.E., Bradshaw, E., et al. (2013). New Data Systems and Products at the Permanent Service for Mean Sea Level. Journal of Coastal Research , 29(3), 493-504. https://doi.org/10.2112/JCOASTRES-D-12-00175.1.

[10] PSMSL. (2018). Tide Gauge Data. Retrieved from https://www.psmsl.org/data/obtaining/.

[11] Cleveland, R.B., Cleveland, W.S., McRae, J.E., and Terpenning, I. (1990). STL: A seasonal-trend decomposition procedure based on loess. Journal of Official Statistics, Vol. 6, pp. 3-73.

[12] Mertz, F., Pujol, M.-I., and Faugère, Y. (2018). Product user manual (Version 4.0). Copernicus Marine Environment Monitoring Service. Retrieved from http://cmems-resources.cls.fr/documents/PUM/CMEMS-SL-PUM-008-032-051.pdf.

[13] Wöppelmann, G., and Marcos, M. (2016). Vertical land motion as a key to understanding sea level change and variability. Reviews of Geophysics, 54(1), 64-92. https://doi.org/10.1002/2015RG000502

[14] Yang, L., and Francis, O. (2019). Historical and future sea level rise rates derived by combining observed and modeled data in the Hawaiian Islands. Coastal Hydraulics Engineering Resilience (CHER) Lab, Department of Civil and Environmental Engineering, University of Hawai'i at Manoa. Mendeley Data, v1 http://dx.doi.org/10.17632/3ks4dtk29v.1.

[15] Yang, L., and Francis, O. (2019). Sea-Level Rise and Vertical Land Motion on the Islands of Oahu and Hawaii, Hawaii. Advances in Space Research, 64(11), 2221-2232, https://doi.org/10.1016/j.asr.2019.08.028.

[16] Blewitt, G., Hammond, W.C., and Kreemer, C. (2018). Harnessing the GPS data explosion for interdisciplinary science. Eos, 99. https://doi.org/https://doi.org/10.1029/2018EO104623.

[17] Hawai'i Climate Change Mitigation and Adaptation Commission. (2017). Hawai'i Sea Level Rise Vulnerability and Adaptation Report. Prepared by Tetra Tech, Inc. and the State of Hawai'i Department of Land and Natural Resources, Office of Conservation and Coastal Lands, under the State of Hawai'i Department of Land and Natural Resources Contract No: 64064.

[18] Sweet, W.V., Kopp, R.E., Weaver, C.P., Obeysekera, J., Horton, R.M., Thieler, E.R., and Chris, Z. (2017). Global and Regional Sea Level Rise Scenarios for the United States (Tech. Rep. NOS CO-OPS 083), Silver Spring, MD: National Oceanic and Atmospheric Administration.

[19] PacIOOS (Pacific Islands Ocean Observing System). (2018). Wave Forecast. Retrieved from http://www.pacioos.hawaii.edu/waves-category/model/, (November 2018).

[20] Anderson, T.R., Fletcher, C.H., Barbee, M.M., Romine, B.M., Lemmo, S., and Delevaux, J.M.S. (2018). Modeling multiple sea level rise stresses reveals up to twice the land at risk compared to strictly passive flooding methods. Scientific Reports, 8(1), 1-14. https://doi.org/10.1038/s41598-018-32658-x.

[21] Vitousek, S., and Fletcher, C.H. (2008). Maximum Annually Recurring Wave Heights in Hawaii. Pacific Science, 62(4), 541-553.

[22] Anderson, T.R., Fletcher, C.H., Barbee, M.M., Frazer, L.N., and Romine, B.M. (2015). Doubling of coastal erosion under rising sea level by mid-century in Hawaii. Natural Hazards, 78(1), 75-103.

[23] Yates, M.L., Le Cozannet, G., and Lenotre, N. (2011). Quantifying errors in long-term coastal erosion and inundation hazard assessments. Journal of Coastal Research, 21(6), 1166-1172. https://doi.org/10.2112/03-0051.1

[24] Davidson-Arnott, R.G.D. (2005). Conceptual model of the effects of sea level rise on sandy coasts. Journal of Coastal Research 21(6):1166-1172.

[25] Fletcher, C.H., Romine, B.M., Genz, A.S., Barbee, M.M., Dyer, M., Anderson, T.R., et al. (2012). National assessment of shoreline change: Historical shoreline change in the Hawaiian Islands. U.S. Geological Survey Open-File Report 2011-1051, 55 p. http://pubs.usgs.gov/of/2011/1051.

[26] Church, J.A., Clark, P.U., Cazenave, A., Gregory, J.M., Jevrejeva, S., Levermann, A., et al. (2013). Sea Level Change. In: Climate Change 2013: The Physical Science Basis. Contribution of Working Group I to the Fifth Assessment Report of the Intergovernmental Panel on Climate Change [Stocker, T.F., Qin, D., Plattner, G.-K., Tignor, M., Allen, S.K., Boschung, J., Nauels, A., Xia, Y., Bex, V.,

and Midgley, P.M. (eds.)]. Cambridge University Press, Cambridge, United Kingdom and New York, NY, USA, pp. 1137-1216, doi:10.1017/CBO9781107415324.026.

[27] Bulter, R. (2012). Re-examination of the Potential for Great Earthquakes along the Aleutian Island Arc with Implications for Tsunamis in Hawaii. Seismological Research Letters, 83(1), 29-38. https://doi.org/https://doi.org/10.1785/gssrl.83.1.29l

[28] Butler, R., Walsh, D., and Richards, K. (2017). Extreme tsunami inundation in Hawai'i from Aleutian--Alaska subduction zone earthquakes. Natural Hazards, 85(3), 1591-1619. https://doi.org/10.1007/s11069-016-2650-0

[29] Cheung, K.F. (2015). Hawaii Tsunami Mapping Project : Data Sources, Procedures, and Products for Extreme Aleutian Events. Final Report Prepared for the Hawaii Emergency Management Agency. Honolulu, HI: University of Hawaii (Personal Communication).

[30] Yamazaki, Y., Kowalik, Z. and Cheung, K.F. (2009). Depth-integrated, non-hydrostatic model for wave breaking and run-up. International Journal for Numerical Methods in Fluids, 61(5), 473-497. https://doi.org/10.1002/fld.1952

[31] Yamazaki, Y., Lay, T., Cheung, K.F., Yue, H., and Kanamori, H. (2011). Modeling near-field tsunami observations to improve finite-fault slip models for the 11 March 2011 Tohoku earthquake. Geophysical Research Letters, 38(20), 18-23. https://doi.org/10.1029/2011GL049130

[32] Cheung, K.F. (2009). Hawaii Tsunami Mapping Project: Data Sources, Procedures, and Products. Final Report for Oahu. Honolulu, HI: University of Hawaii.

[33] Cheung, K.F. (2011). Hawaii Tsunami Mapping Project: Data Sources, Procedures, and Products. Final Report for the County of Hawaii. Honolulu, HI: University of Hawaii

[34] Cheung, K.F. (2012). Hawaii Tsunami Mapping Project: Data Sources, Procedures, and Products. Final Report for the County of Maui. Honolulu, HI: University of Hawaii.

[35] Cheung, K.F. (2013). Hawaii Tsunami Mapping Project: Data Sources, Procedures, and Products. Final Report for the County of Kauai. Honolulu, HI: University of Hawaii.

[36] Zachry, B.C., Booth, W.J., Rhome, J.R., and Sharon, T.M. (2015). A National View of Storm Surge Risk and Inundation. *Weather, Climate, and Society* , 7(2), 109-117. https://doi.org/10.1175/WCAS-D-14-00049.1

[37] Togia, H., Francis, O., Kim, K., and Zhang, G. (2019). Segment-based approach for assessing hazard risk of coastal highways in Hawai'i. Transportation Research Record: Journal of the Transportation Research Board, Sage Publications Inc. https://doi.org/10.1177/0361198118821679.

[38] Francis, O., Kim, K., and Pant, P. (2019). Stakeholder assessment of coastal risks and mitigation strategies. Ocean & Coastal Management, 179, https://doi.org/10.1016/j.ocecoaman.2019.104844.

[39] Francis, O. (2018). Coastal Adaptive Design and Adaptation. In: Climate-Resilient Infrastructure: Adaptive Design and Risk Management, MOP 140, Committee on Adaptation to a Changing Climate, American Society of Civil Engineers, [Ayyub, B.M., Walker, D., Wright, R.N., Vinson, T., and Medina, M. (eds.)]. ISBN 9780784415191, https:// doi.org/10.1061/9780784415191.